COLLEGE PLACEMENT MATH
in 20 Minutes a Day

Other Titles of Interest from LearningExpress

COLLEGE PLACEMENT MATH
in 20 Minutes a Day

Catherine V. Jeremko
and
Colleen M. Schultz

NEW YORK

CONTRIBUTORS

Catherine V. Jeremko is a certified secondary mathematics teacher in New York State. She is the author of mathematics test preparation and review materials, as well as a manuscript reviewer for several publishing companies. She currently teaches seventh grade mathematics at Vestal Middle School in Vestal, New York. Ms. Jeremko is also a teacher trainer for both mathematical pedagogy and the use of technology in the classroom. She resides in Apalachin, New York, with her three daughters.

Colleen M. Schultz is certified in both secondary mathematics and elementary education in New York State. She is the author of mathematics test preparation and review materials, as well as a manuscript reviewer for several publishing companies. She currently teaches eighth grade mathematics at Vestal Middle School in Vestal, New York, where she recently served as a teacher mentor for the Vestal School District. Mrs. Schultz is also a teacher trainer for both mathematical pedagogy and the use of technology in the classroom. She resides in Binghamton, New York, with her husband and three children.

Scott Meltzer is a National Board–certified middle school math teacher. His 10 years of experience as an educator have included work at all levels, elementary through college, and he has served as both a teacher and a coach, providing professional development for colleagues. A New York native, he now lives in Seattle with his wife and two sons.

CONTENTS ▶

HOW TO USE THIS BOOK ix

PRETEST 1

LESSON 1 Number Sense 21
For the COMPASS, ASSET, and ACCUPLACER

LESSON 2 Integers and Operations 29
For the COMPASS, ASSET, and ACCUPLACER

LESSON 3 Rational Numbers 35
For the COMPASS, ASSET, and ACCUPLACER

LESSON 4 Ratio and Proportion 45
For the COMPASS, ASSET, and ACCUPLACER

LESSON 5 Percent and Applications 51
For the COMPASS and ACCUPLACER

LESSON 6 Exponents and Square Roots 59
For the COMPASS, ASSET, and ACCUPLACER

LESSON 7 Algebraic Expressions 69
For the COMPASS, ASSET, and ACCUPLACER

CONTENTS

LESSON 8   **Linear Equations and Word Problems**   75
For the COMPASS, ASSET, and ACCUPLACER

LESSON 9   Inequalities   83
For the ASSET and ACCUPLACER

LESSON 10   Basic Geometry   91
For the COMPASS, ASSET, and ACCUPLACER

LESSON 11   Measurement   101
For the COMPASS, ASSET, and ACCUPLACER

LESSON 12   Coordinate Geometry   115
For the COMPASS, ASSET, and ACCUPLACER

LESSON 13   Systems of Equations   125
For the COMPASS, ASSET, and ACCUPLACER

LESSON 14   Polynomials and Factoring   133
For the COMPASS, ASSET, and ACCUPLACER

LESSON 15   Rational Expressions and Equations (Including Quadratic Equations)   141
For the COMPASS, ASSET, and ACCUPLACER

LESSON 16   Functions   149
For the COMPASS, ASSET, and ACCUPLACER

LESSON 17   Exponential Functions and Logarithms   155
For the COMPASS, ASSET, and ACCUPLACER

LESSON 18   Trigonometry   163
For the COMPASS, ASSET, and ACCUPLACER

LESSON 19   Trigonometric Identities and Equations   173
For the COMPASS and ACCUPLACER

LESSON 20   Complex Numbers, Sequences and Series, and Matrices   181
For the COMPASS, ASSET, and ACCUPLACER

POSTTEST   193

ADDITIONAL ONLINE PRACTICE   213

HOW TO USE THIS BOOK ▶

Colleges use your scores on college placement tests like the COMPASS, ASSET, and ACCUPLACER, along with your high school transcript, to help place you in the appropriate math courses or course levels. Placing into higher-level math courses can save you time and money as you navigate your way through college.

College Placement Math in 20 Minutes a Day is organized into small, manageable lessons—lessons you can master in 20 minutes a day or easily break up into 20-minute chunks. The first page of each lesson lets you know which of the three main placement exams—COMPASS, ASSET, and ACCUPLACER—tests the math featured in that lesson. Each lesson presents a topic found on the math placement exams, one step at a time.

The lessons teach by example—rather than by theory—so you have plenty of opportunities for successful learning. You'll learn by understanding, not by memorization. Each new topic is introduced with practical, easy-to-follow examples, which walk you through the solution process. Most topics are then reinforced by sample questions for you to try on your own.

You'll also find lots of valuable memory hooks and shortcuts to help you retain what you're learning. Practice question sets, scattered throughout each lesson, typically begin with easy questions to help build your confidence. As the lessons progress, these easier questions are interspersed with the more challenging ones, so that even readers who are having trouble can successfully complete many of the questions. A little success goes a long way!

You should start by taking the pretest that begins on page 1. The pretest will tell you which lessons you should really concentrate on. At the end of the book, you'll find a posttest that will show you how much you've improved.

Keep in mind that there is no set passing score on any section of the COMPASS, ASSET, or ACCUPLACER—institutions use the tests to see in which areas you need extra instruction and in which areas you already have excelled. Every institution uses the scores differently.

This is a workbook, and as such, it's meant to be written in. Unless you checked it out from a library or borrowed it from a friend, write all over it! Get actively involved in doing each math problem; mark up the chapters boldly. You may even want to keep extra paper available, because sometimes you could end up using two or three pages of scratch paper for one problem—and that's fine!

Make a Commitment

You've got to take your math preparation further than simply reading this book. Improving your math skills takes time and effort on your part. You have to make the commitment. You have to carve time out of your busy schedule. You have to decide that improving your skills—improving your chances of doing well in college math—is a priority for you.

If you're ready to make that commitment, this book will help you. Since each of its 20 lessons is designed to be completed in 20 to 60 minutes, you can build a firm math foundation in just one month, conscientiously working through the lessons for 20 minutes a day, five or six days a week. If you follow the tips for continuing to improve your skills and do each of the exercises, you'll build an even stronger foundation.

Use this book to its fullest extent—as a self-teaching guide and then as a reference resource—to get the fullest benefit.

Now that you're armed with a positive math attitude, it's time to dig into the pretest. Go for it!

PRETEST

Before you begin the first lesson to prepare you for your upcoming test, it's a good idea to find out how much you already know and how much you need to learn. Take the pretest in this chapter, which includes 50 multiple-choice questions covering the topics in this book. While this pretest can't cover every skill reviewed in this book, your performance on the pretest will give you a good indication of your strengths and weaknesses.

If you score high on the pretest, you have a good foundation and should be able to work your way through the book quickly. If you score low on the pretest, don't despair. This book will take you through the concepts step-by-step. If you get a low score, you may need more than 20 minutes a day to work through a lesson. However, this is a self-paced program, so you can spend as much time on a lesson as you need. You decide when you fully comprehend the lesson and are ready to go on to the next one. You will also notice that the degree of difficulty will increase through the pretest, as the skills get more complex as the lessons progress. You may find you need to spend more time on those advanced topics.

Take as much time as you need to complete the pretest. When you have finished, check your answers in the answer key at the end of the pretest. Each answer also tells you which lesson of this book teaches you about the skills needed for that question.

1.	ⓐ	ⓑ	ⓒ	ⓓ
2.	ⓐ	ⓑ	ⓒ	ⓓ
3.	ⓐ	ⓑ	ⓒ	ⓓ
4.	ⓐ	ⓑ	ⓒ	ⓓ
5.	ⓐ	ⓑ	ⓒ	ⓓ
6.	ⓐ	ⓑ	ⓒ	ⓓ
7.	ⓐ	ⓑ	ⓒ	ⓓ
8.	ⓐ	ⓑ	ⓒ	ⓓ
9.	ⓐ	ⓑ	ⓒ	ⓓ
10.	ⓐ	ⓑ	ⓒ	ⓓ
11.	ⓐ	ⓑ	ⓒ	ⓓ
12.	ⓐ	ⓑ	ⓒ	ⓓ
13.	ⓐ	ⓑ	ⓒ	ⓓ
14.	ⓐ	ⓑ	ⓒ	ⓓ
15.	ⓐ	ⓑ	ⓒ	ⓓ
16.	ⓐ	ⓑ	ⓒ	ⓓ
17.	ⓐ	ⓑ	ⓒ	ⓓ

18.	ⓐ	ⓑ	ⓒ	ⓓ
19.	ⓐ	ⓑ	ⓒ	ⓓ
20.	ⓐ	ⓑ	ⓒ	ⓓ
21.	ⓐ	ⓑ	ⓒ	ⓓ
22.	ⓐ	ⓑ	ⓒ	ⓓ
23.	ⓐ	ⓑ	ⓒ	ⓓ
24.	ⓐ	ⓑ	ⓒ	ⓓ
25.	ⓐ	ⓑ	ⓒ	ⓓ
26.	ⓐ	ⓑ	ⓒ	ⓓ
27.	ⓐ	ⓑ	ⓒ	ⓓ
28.	ⓐ	ⓑ	ⓒ	ⓓ
29.	ⓐ	ⓑ	ⓒ	ⓓ
30.	ⓐ	ⓑ	ⓒ	ⓓ
31.	ⓐ	ⓑ	ⓒ	ⓓ
32.	ⓐ	ⓑ	ⓒ	ⓓ
33.	ⓐ	ⓑ	ⓒ	ⓓ
34.	ⓐ	ⓑ	ⓒ	ⓓ

35.	ⓐ	ⓑ	ⓒ	ⓓ
36.	ⓐ	ⓑ	ⓒ	ⓓ
37.	ⓐ	ⓑ	ⓒ	ⓓ
38.	ⓐ	ⓑ	ⓒ	ⓓ
39.	ⓐ	ⓑ	ⓒ	ⓓ
40.	ⓐ	ⓑ	ⓒ	ⓓ
41.	ⓐ	ⓑ	ⓒ	ⓓ
42.	ⓐ	ⓑ	ⓒ	ⓓ
43.	ⓐ	ⓑ	ⓒ	ⓓ
44.	ⓐ	ⓑ	ⓒ	ⓓ
45.	ⓐ	ⓑ	ⓒ	ⓓ
46.	ⓐ	ⓑ	ⓒ	ⓓ
47.	ⓐ	ⓑ	ⓒ	ⓓ
48.	ⓐ	ⓑ	ⓒ	ⓓ
49.	ⓐ	ⓑ	ⓒ	ⓓ
50.	ⓐ	ⓑ	ⓒ	ⓓ

Pretest

1. What composite number is greater than 38 but less than 50?
   a. 41
   b. 47
   c. 49
   d. 55

2. Solve: $3 \times 7 - 4(8 - 5) + 2$.
   a. 7
   b. 11
   c. 23
   d. 53

3. What is the value of $\frac{9!}{7!}$?
   a. 1.2857
   b. 2
   c. 72
   d. 51,840

4. Find the median of the following set of numbers: 9, 8, 18, 11, 15, 10, 13, 15, and 9.
   a. 9
   b. 10
   c. 11
   d. 15

5. Simplify: $-242 \div (-6 - 5) \times 2$.
   a. 11
   b. 22
   c. 44
   d. −44

6. Simplify: $20 \div 4 \times -5 + |-18 + 6|$.
   a. 13
   b. −13
   c. 37
   d. −37

7. What is $\frac{1}{8}$ of 24?
   a. 2
   b. 3
   c. $\frac{1}{3}$
   d. 4

8. Stephanie purchases one dress for $19.99, one dress for $26.99, a pair of shoes for $35.99, and two pairs of stockings for $3.99 each. She uses the $100 bill her grandmother sent her for Christmas to pay for the items. How much change does she go home with?
   a. $5.05
   b. $13.04
   c. $9.95
   d. $9.05

9. Jalapeño peppers cost 3 for $0.99. At that rate, how much will 27 peppers cost, to the nearest penny?
   a. $8.91
   b. $26.73
   c. $81.00
   d. $81.81

10. The scale for a road map indicates that $\frac{1}{2}$ inch equals 15 miles. The distance on U.S. Route 1 between Stafford, VA, and Blantons, VA, is $1\frac{3}{8}$ inches. Approximately how many miles are there between Stafford and Blantons when traveling on U.S. Route 1?
   a. 0.05
   b. 20.63
   c. 30
   d. 41.25

11. Two numbers are in the ratio 7:2. The sum of the two numbers is 108. What is the smaller number?
   a. 12
   b. 24
   c. 84
   d. 96

12. If Suzanne sold her one-of-a-kind designer dress, which she had purchased for $15,000, for $17,250, what percentage profit did she make?
   a. 15%
   b. 10%
   c. 7.5%
   d. 20%

13. Write $\frac{7}{56}$ as a percentage.
   a. 7.56%
   b. 17.56%
   c. 12.5%
   d. 25%

14. Hannah bought $\frac{7}{8}$ of a pound of chocolates and ate $\frac{1}{4}$ of a pound. What percentage of a pound was left?
   a. 15%
   b. 75%
   c. 62.5%
   d. 25%

15. Find the product: $(3^2)(3^3)$.
   a. 81
   b. 243
   c. 729
   d. 59,049

16. Simplify: $\sqrt{28} + \sqrt{63}$.
   a. $\sqrt{91}$
   b. $2\sqrt{7} + 3\sqrt{7}$
   c. $5\sqrt{7}$
   d. $13\sqrt{7}$

17. Write 0.00000272338 in scientific notation.
   a. 2.72338×10^{-5}
   b. 272.338×10^{-7}
   c. 2.72338×10^{-6}
   d. $272,338 \times 10^{-11}$

18. Simplify the expression: $3(5x - 4) - 4(4x + 5)$.
   a. $x - 32$
   b. $-x + 8$
   c. $31x - 8$
   d. $-x - 32$

19. Simplify the expression: $3(ab - a) - 3(b - a) - 6ab$.
   a. $-3ab - 3b$
   b. $3ab + 3b$
   c. $-3ab + 3b$
   d. $-3ab - 6ax - 3bx$

20. Perform the indicated operations and combine like terms: $(4x^2 + 8x - 7) - (5x^2 - 6x + 2)$.
   a. $-x^2 + 14x + 9$
   b. $x^2 + 14x + 9$
   c. $-x^2 + 14x - 9$
   d. $-x^2 + 2x - 5$

21. Solve the following equation for x: $\frac{5x}{4} = 6 + 2x$.
   a. 8
   b. −8
   c. 3
   d. 2

22. If $a = 8b$, and $48b = 4c$, then what is c in terms of a?
   a. $\frac{1}{4}a$
   b. $2a$
   c. $\frac{1}{8}a$
   d. $\frac{3}{2}a$

23. Matthew purchases medallions wholesale from the manufacturer for $3.95 each. He then sells them for $6.95 each. His fixed costs each month are $5,000 for rent and utilities. If his average monthly sales are n (number of medallions), which equation can be used to calculate his average monthly profit (P)?

a. $P = n(6.95 - 3.95) - 5,000$
b. $P = n(5,000) + (6.95 - 3.95)$
c. $P = n(6.95 - 3.95) + 5,000$
d. $P = n(6.95 - 3.95)$

24. Solve $4x \leq -\frac{2}{3}$.
a. $x \leq -\frac{8}{3}$
b. $x \leq \frac{1}{6}$
c. $x \geq -\frac{1}{6}$
d. $x \leq -\frac{1}{6}$

25. The number line shown is a graph of the solution set of which of the following linear inequalities?

a. $-\frac{2}{3}x < 12$
b. $-\frac{2}{3}x \leq 12$
c. $-\frac{2}{3}x \geq 12$
d. $-\frac{2}{3}x > 12$

26. Two angles of a triangle measure 35° and 115°. What is the measure of the exterior angle adjacent to the third angle?
a. 30°
b. 50°
c. 150°
d. 180°

27. Which of the following figures does/do NOT have line(s) of symmetry?

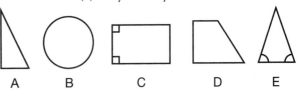

a. D only
b. A and D
c. B, C, and D
d. A, D, and E

28. In the following figure, given that line a is parallel to line b, and angle 8 = 32°, what are the measurements for angles 5, 4, and 2?

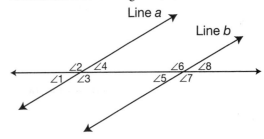

a. angle 5 = 32°, angle 4 = 148°, angle 2 = 32°
b. angle 5 = 148°, angle 4 = 148°, angle 2 = 32°
c. angle 5 = 32°, angle 4 = 32°, angle 2 = 148°
d. angle 5 = 32°, angle 4 = 148°, angle 2 = 148°

29. If the length of one side of a cube is 5 mm, what is the cube's surface area?
a. 30 mm^2
b. 125 mm^3
c. 150 mm^2
d. 156 mm^2

30. If a rectangle is 12 m high and 24 m wide, what is its area?
a. 144 m^2
b. 288 m^2
c. 72 m^2
d. 576 m^2

31. Find the length of the hypotenuse of a right triangle with leg length of 9 units and 12 units.
   a. 8 units
   b. 15 units
   c. 21 units
   d. 225 units

32. What is the slope of the line on a coordinate graph that passes through the ordered pairs $(5,-2)$ and $(3,1)$?
   a. $-\frac{3}{2}$
   b. $\frac{3}{8}$
   c. $-\frac{2}{3}$
   d. $\frac{3}{2}$

33. Find the distance between the points $(-3,2)$ and $(-2,3)$.
   a. 2
   b. $\sqrt{50}$
   c. $\sqrt{2}$
   d. $\sqrt{10}$

34. What is the solution to this system of equations?

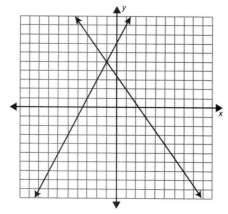

   a. $(5,-1)$
   b. $(0,3.5)$
   c. $(-1,5)$
   d. $(1,5)$

35. Which point is in the solution set of the following inequality system?

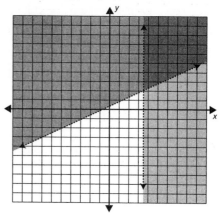

   a. $(3.5,2)$
   b. $(5,6)$
   c. $(2,3.5)$
   d. $(6,2)$

36. Solve the system algebraically: $y = x - 4$ and $2y + 4x = 19$.
   a. $(4.5,0.5)$
   b. $(4,0)$
   c. $(3.83,-0.17)$
   d. $(7,3)$

37. Factor the following polynomial expression completely: $3x^3 + 6x^2 + 3x$.
   a. $3x(x + 1)(x + 1)$
   b. $3x(x + 1)(x - 1)$
   c. $(3x^2 + 3x)(x + 1)$
   d. $3x$

38. Multiply: $(x^2 + 3x - 2)^2$.
   a. $x^4 + 3x^3 + 5x^2 - 12x + 4$
   b. $x^4 + 6x^3 + 5x^2 - 12x + 4$
   c. $x^4 + 6x^3 + 5x^2 + 12x + 4$
   d. $x^4 + 6x^3 + 9x^2 - 12x + 4$

39. Simplify the following expression: $(\frac{5x^3}{4})(\frac{12}{20x})$.
   a. $\frac{6x^3}{8x}$
   b. $\frac{3x^4}{4}$
   c. $\frac{3x^2}{8}$
   d. $\frac{3x^2}{4}$

40. Simplify the following expression: $\frac{(x^2 + 3x - 10)}{x + 3} \div \frac{x + 5}{(x^2 - 2x - 15)}$.

a. $x^2 + 3x + 10$

b. $\frac{(x + 3)^2(x - 5)}{(x + 5)^2(x - 2)}$

c. $x^2 - 7x + 10$

d. $\frac{(x + 5)^2(x - 2)}{(x + 3)^2(x - 5)}$

41. Evaluate $f(12)$ when $f(x) = -3x + 10$.

a. 7

b. 46

c. 19

d. −26

42. Evaluate $f \circ g(-4)$, when $f(x) = x^3 - 1$ and $g(x) = x + 1$.

a. −126

b. 26

c. −28

d. −10

43. Given $f(x) = -5(\frac{1}{4})^x - 3$, evaluate $f(-1)$ and give the ordered pair.

a. $(-1, -23)$

b. $(-1, 17)$

c. $(-1, -20)$

d. $(-1, -\frac{17}{4})$

44. Which of the following graphs depicts an exponential function?

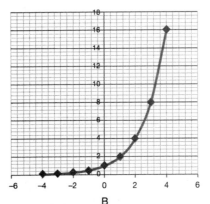

B

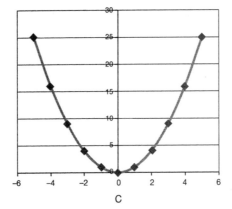

C

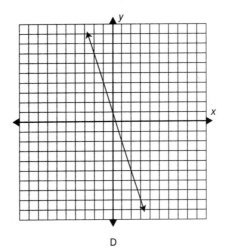

D

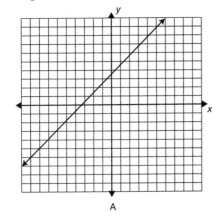

A

a. A

b. B

c. C

d. D

45. Solve for n: $\log_3 n - \frac{\log_3 1}{4} = \log_3 36$.
- **a.** 144
- **b.** $\log_3 144$
- **c.** 9
- **d.** $\log_3 9$

46. Solve for x: $4^x = 24$. If needed, round to the nearest hundredth.
- **a.** $x = 6$
- **b.** $x = 96$
- **c.** $x \approx 0.44$
- **d.** $x \approx 2.29$

47. If $\sin \alpha = \frac{17}{21}$ and $\cos \alpha = \frac{15}{21}$, $\tan \alpha = ?$
- **a.** $\frac{15}{17}$
- **b.** $\frac{17}{15}$
- **c.** $\frac{21}{15}$
- **d.** $\frac{21}{17}$

48. Find all values satisfying the equation $\cos^2 x + 3\cos x = -2$ in the interval $0° \leq x < 360°$.
- **a.** $0°$
- **b.** $90°$
- **c.** $0°, 180°$
- **d.** $180°$

49. Simplify $\frac{\sqrt{-32}}{\sqrt{-64}}$.
- **a.** $\frac{1}{2}$
- **b.** $\frac{1}{2}i$
- **c.** $\frac{\sqrt{2}}{2}$
- **d.** $\frac{\sqrt{2}}{8}$

50. Evaluate $\sum_{n=1}^{20} -2n + 5$.
- **a.** -320
- **b.** -35
- **c.** 4
- **d.** -7.5

Answers

1. c. (Lesson 1) Composite numbers have factors in addition to 1 and themselves; 1 is neither prime nor composite. The number 7 is a factor of 49 in addition to 1 and 49; therefore, 49, answer choice **c**, is a composite number. Choices **a** and **b** are incorrect—41 and 47 are prime numbers. Choice **d** is incorrect—55 is a composite number, but larger than 50.

2. b. (Lesson 1) Choice **b** is correct. Remember PEMDAS ("Please Excuse My Dear Aunt Sally"): perform operations within parentheses first. Working from left to right, calculate all multiplication and division next. Finally, work from left to right to calculate all additions and subtractions.
$8 - 5 = 3$
$3 \times 7 = 21$
$-4 \times 3 = -12$
$21 - 12 + 2 = 11$
Choice **a** is incorrect because $3 \times 7 - 4(8 - 5) + 2 = 11$, not 7. Choice **c** is incorrect because $3 \times 7 - 4(8 - 5) + 2 = 11$, not 23. Choice **d** is incorrect and could be the result of subtracting $7 - 4$ as an incorrect first step in the problem.

3. c. (Lesson 1) Solve as follows: $\frac{9!}{7!} =$
$$\frac{\cancel{1} \times \cancel{2} \times \cancel{3} \times \cancel{4} \times \cancel{5} \times \cancel{6} \times \cancel{7} \times 8 \times 9}{\cancel{1} \times \cancel{2} \times \cancel{3} \times \cancel{4} \times \cancel{5} \times \cancel{6} \times \cancel{7}}$$
$= 8 \times 9 = 72$. Factorials are simply the product of all of the whole numbers from 1 to n. $n! = 1 \times 2 \times 3 \times \ldots \times n$. Thus, $\frac{9!}{7!} = 72$. Choice **a** is the result of simply dividing 9 by 7. Choice **b** is the result of subtracting $9 - 7$. Choice **d** is incorrect because it is the result of dividing 9 factorial by 7, not 7 factorial.

4. c. (Lesson 1) The median of this set of numbers is 11. The median is the middle value in a set of numbers. Rewrite the set of numbers in numerical order, to find the middle value in the ordered set: 8, 9, 9, 10, 11, 13, 15, 15, 18. The middle value is 11. Choice **a** is incorrect and is one of the modes of the set, the number that appears most often. Choice **b** is the range of the set, or the highest minus the lowest value. Choice **d** is incorrect because the median of the set is found after writing the set in numerical order.

5. c. (Lessons 1 and 2) To simplify $-242 \div (-6 - 5) \times 2$, first perform the subtraction in parentheses to get $-242 \div -11 \times 2$. Next, divide to get $22 \times 2 = 44$. If you chose answer **a**, you most likely did multiplication before division. These operations should be done from left to right. If you chose answer **b**, you most likely forgot to multiply by 2 as the final step. If you chose answer **d**, you incorrectly simplified $-242 \div -11$ as negative 22, when it is positive 22 because a negative factor times a negative factor is a positive product.

6. b. (Lessons 1 and 2) To simplify $20 \div 4 \times -5 + |-12|$, first perform the addition in the absolute value symbol, which is a grouping symbol, to get $20 \div 4 \times -5 + |-12|$. Absolute value is always positive: $20 \div 4 \times -5 + 12$. Divide next to get $5 \times -5 + 12$. Multiply and then add to get -13. If you chose answer **a**, you most likely thought that $5 \times -5 = 25$, and also treated the absolute value symbols like parentheses. If you chose answer **c**, you incorrectly simplified $5 \times -5 = 25$. If you chose answer **d**, you correctly simplified $5 \times -5 = -25$, but treated the absolute value symbols like parentheses.

7. b. (Lesson 3) To find the answer, simply divide 24 by the denominator of the fraction, 8. Then, $24 \div 8 = 3$. Choice **a** is incorrect—it multiplies 24 by $\frac{1}{12}$, not $\frac{1}{8}$. Choice **c** is incorrect—it multiplies 8 by $\frac{1}{24}$, not $\frac{1}{8}$ by 24. Choice **d** is incorrect—it multiplies 24 by $\frac{1}{6}$, not $\frac{1}{8}$.

8. d. (Lesson 3) Rounding each of the purchases to the nearest dollar, Stephanie purchased $20 + 27 + 36 + 8 = 91$. Her change will be approximately $9.00, making answer choice **d** the best choice. Checking: $19.99 + 26.99 + 35.99 + 3.99 + 3.99 = 90.95$; $100.00 - 90.95 = 9.05$. Choice **a** is incorrect: $19.99 + 26.99 + 35.99 + 3.99 + 3.99 = 90.95$; $100.00 - 90.95 = 9.05$, not $5.05. Choice **b** is incorrect. It neglects to add *two* pairs of stockings for $3.99 each, not just one. Choice **c** is incorrect. It incorrectly subtracts 90.95 from 100: $100.00 - 90.95 = 9.05$, not $9.95.

9. a. (Lesson 4) Set up a proportion of $\frac{jalapeño}{dollars} = \frac{jalapeño}{dollars}$. The proportion is $\frac{3}{0.99} = \frac{27}{x}$. Solve to get the cost equal to $8.91. If you chose answer **b**, you mostly likely treated the cost as $0.99 for one jalapeño. If your answer was choice **c**, you incorrectly multiplied 27×3, instead of $27 \times 0.99 \div 3$. If your answer was choice **d**, you found the unit cost of one jalapeño.

10. d. (Lesson 4) Change the fractions into decimal equivalents. Set up a proportion using the scale as 0.5 inch = 15 miles. The proportion is $\frac{0.5}{15} = \frac{1.375}{x}$. Cross multiply to get $0.5x = 20.625$. Divide both sides by 0.5 and $x = 41.25$ miles. If your answer was choice **a**, you set up the proportion incorrectly. If your answer was choice **b**, most likely when you cross multiplied you divided by 2 instead of dividing by $\frac{1}{2}$. If your answer was choice **c**, you used 1 inch instead of $1\frac{3}{8}$ inch for the map distance.

11. b. (Lesson 4) The smaller number is two parts out of a total of $2 + 7 = 9$. Set up the proportion $\frac{2}{9} = \frac{x}{108}$. Cross multiply to get $216 = 9x$. Divide both sides by 9 to get $x = 24$. If you chose answer **a**, you most likely solved to find one part out of 9, which is 12. If your choice was answer **c**, you found the larger of the two numbers. If your choice was answer **d**, you may have added half the value of x to the value of the larger number, 84.

12. a. (Lesson 5) To calculate percentage profit, first find the amount of profit. Profit = sale price (17,250) – original purchase price (15,000) = 2,250. Now, divide 2,250 (the profit) by 15,000 (the original purchase price). The result is 0.15. To convert this to a percent, multiply by 100: $0.15 \times 100 = 15\%$. Choice **b** is incorrect. A 10% percentage profit would be $(0.10)(15,000) = 1,500$, and $1,500 + 15,000 = \$16,500$, not \$17,250. Choice **c** is also incorrect, and could be the result of a calculation error. Choice **d** is incorrect. A 20% profit would be $(0.20)(15,000) = 3,000$, and $3,000 + 15,000 = \$18,000$, not \$17,250.

13. c. (Lesson 5) To convert a fraction to a percentage, divide the numerator by the denominator: $7 \div 56 = 0.125$. Then multiply by 100 to get the percentage = 12.5%. Choice **a** is incorrect and is the result of placing the numerator and denominator in decimal form. Choice **b** is incorrect and could also be the result of placing the numbers from the fraction within a decimal. Choice **d** is incorrect and is double the value of the correct answer.

14. c. (Lesson 5) First convert the fractions so that you have a common denominator. To make 8 the denominator of $\frac{1}{4}$, multiply the fraction by $\frac{2}{2}$:
$$\frac{1}{4} \times \frac{2}{2} = \frac{2}{8}$$
Subtract:
$$\frac{7}{8} - \frac{2}{8} = \frac{5}{8}$$
Now convert the fraction to a percent by dividing the numerator by the denominator, and then multiplying the decimal by 100. The answer becomes:
$$\frac{5}{8} = 0.625 = 62.5\%$$

15. b. (Lesson 6) When you multiply two powers with the same base, keep the base and add the exponents. In this case, the base is 3: $(3^2)(3^3) = 3^2 \times 3^3 = 3^{2+3} = 3^5 = 3 \times 3 \times 3 \times 3 \times 3 = 243$. If your answer was choice **a**, you solved for 3^4 instead of 3^5. If you chose answer **c**, you most likely multiplied the exponents to get $2 \times 3 = 6$, and $3^6 = 729$. If you chose answer **d**, you multiplied the bases to get 9^5, which is 59,049.

16. c. (Lesson 6) To simplify this sum, first simplify each radical separately:
$$\sqrt{28} = \sqrt{2^2 \times 7} = 2\sqrt{7}$$
$$\sqrt{63} = \sqrt{3^2 \times 7} = 3\sqrt{7}$$
These simplified radicals have the same radicand, so you can add them to get $2\sqrt{7} + 3\sqrt{7} = 5\sqrt{7}$. If your answer was choice **a**, you incorrectly added the radicands. If your answer was choice **b**, you did not fully simplify. If your answer was choice **d**, you did not simplify correctly: $\sqrt{28} \neq 4\sqrt{7}$ and $\sqrt{63} \neq 9\sqrt{7}$.

17. c. (Lesson 6) To change a number from standard form to scientific notation, you must change the form of the number to be a number greater than or equal to 1 and less than 10, times the appropriate power of 10. For this number you must move the decimal point six places to the right, to get 2.72338×10^{-6}. The value of the exponent was -6 because the decimal point was moved six places, and the original number was a very small one, making the exponent negative. If your answer was choice **a**, you counted the number of leading zeros, instead of counting the number of places you moved the decimal point. If you chose answer **b**, you made a number that was 272, not 2.7, and counted from the far right as to the number of places moved instead of from the decimal point. If your answer was choice **d**, you moved the decimal point from where it was originally to the far right.

18. d. (Lesson 7) Use the distributive property first: $3(5x - 4) = 15x - 12$, and $4(4x + 5) = 16x + 20$. Then regroup and combine like terms: $15x - 12 - 16x - 20 = 15x - 16x - 12 - 20 = -x - 32$, which is answer choice **d**. Choice **a** is incorrect—it leaves out the negative sign before the x. Choice **b** is incorrect: it improperly combines -12 and -20 to equal 8, not -32. Choice **c** is incorrect—it adds all the terms instead of taking into account the negative signs throughout.

19. a. (Lesson 7) First use the distributive property to eliminate the parentheses: $3ab - 3a - 3b + 3a - 6ab$. The second 3 must be distributed as a -3 since the subtraction sign is in front of it. Then combine like terms to get answer choice **a**, $-3ab - 3b$. Choice **b** is incorrect—it neglects to subtract $-6ab$ from the final expression and has the wrong sign on the $3b$ term. Choice **c** is incorrect—it leaves off the negative sign before $3b$. Choice **d** is incorrect—there are no x's in the expression.

20. c. (Lesson 7) When performing subtraction of polynomials, add the opposite of the second polynomial. $4x^2 - 5x^2 = -x^2$; $8x + 6x = 14x$; and $-7 - 2 = -9$, to yield answer choice **c**, $-x^2 + 14x - 9$. Choice **a** is incorrect: $-7 - 2 = -9$, not 9. Choice **b** is incorrect because it neglects to make the term x^2 negative. Choice **d** is incorrect because it neglects to distribute the negative sign over to all of the terms in the second polynomial.

21. b. (Lesson 8) First, multiply each side of the equation by 4 to clear the denominator. The equation becomes:

$$5x = 24 + 8x$$

$-5x$	$-5x$	Subtract $5x$ from both sides (the opposite of $+5x$).

$$0 = 24 + 3x$$

$-24 =$	$3x$	Subtract 24 from both sides (the opposite of $+24$).
$\frac{-24}{3} =$	$\frac{3x}{3}$	Divide both sides by 3 (the opposite of multiply).
$-8 =$	x	The variable is isolated, so the equation is solved.

If your answer was choice **a**, you may have made an integer arithmetic mistake, or forgotten to multiply $2x$ times 4. Every term needs to be multiplied by 4 to clear the denominator. If your answer was choice **c**, you multiplied 4 and $5x$, when the 4's should have canceled for this term. If you chose answer **d**, you ignored the denominator of 4.

22. d. (Lesson 8) Take the first equation and multiply both sides by 6 so that the **b** term has a coefficient of 48:

$6(a) = 6(8b)$

$6a = 48b$

The second equation says that $48b = 4c$. Substitute in $6a$ for the term $48b$ to get $6a = 4c$. Now, isolate c:

$\frac{6a}{4} = \frac{4c}{4}$

$c = \frac{6a}{4} = \frac{3}{2}a$

If you chose answer **a**, you simply set $a = 4c$, and solved for **c**. If your answer was choice **b**, you multiplied the first equation by 8 instead of 6. If your answer was choice **c**, you solved the equation for the variable b instead of solving for the variable c.

23. a. (Lesson 8) The profit for each medallion is the amount charged minus the wholesale cost. Since n medallions are sold, the profit from the medallions is $n(6.95 - 3.95)$. The final profit, however, is this amount minus the fixed monthly costs of $5,000. The monthly profit is therefore $P = n(6.95 - 3.95) - 5,000$. If your answer was choice **b**, you multiplied the number of medallions by the fixed monthly cost. If you chose answer **c**, you added the fixed cost to the medallion profit instead of subtracting. If your answer was choice **d**, you did not account for the monthly fixed cost.

24. d. (Lesson 9) You solve linear inequalities, for the most part, the same way as you do linear equations. The answer in this format reflects that it is the set of all x values such that x is a real number that is less than, less than or equal to, greater than, or greater than or equal to a real number. So for $4x \le -\frac{2}{3}$, divide both sides of the equation by 4 to get $x \le -\frac{2}{3} \div \frac{4}{1} = x \le -\frac{2}{3} \times \frac{1}{4} = x \le -\frac{2}{12} = -\frac{1}{6}$. Or alternatively, multiply both sides of the equation by 3 to get $12x \le -2 = x \le -\frac{2}{12} = x \le -\frac{1}{6}$. Choice **a** is incorrect. This answer multiplies the other side by 4 instead of dividing by 4. Choice **b** is incorrect—it leaves the negative sign off $-\frac{1}{6}$. Choice **c** is incorrect. When solving inequalities, you flip the sign only if you multiply or divide through by a negative.

25. c. (Lesson 9) The number line depicts a solution set that includes −18 and all values less than −18. The closed circle, or closed interval notation, indicates a less than or equal to ($\le$) or greater than or equal to ($\ge$) inequality in which the endpoint is included in the solution set. An open circle, in contrast, indicates an open interval, or one in which the solution set does not contain the value at the open circle. The key to solving inequalities correctly is remembering that when you multiply or divide both sides by a negative, you must reverse the inequality sign.

$-\frac{2}{3}x \ge 12 \Rightarrow (-3)(\frac{-2}{3}x) \ge (-3)12 \Rightarrow 2x \le -36 \Rightarrow x \le -18$

Choices **a** and **d** are incorrect because the closed circle notation indicates that the solution set includes −18, and the graphs to these answer choices would have an open circle. Choice **b** is incorrect because the inequality sign indicates less than or equal to, and when multiplying by a negative, the inequality sign should be reversed.

26. c. (Lesson 10) The sum of the interior angles of a triangle is 180°. Because 35 + 115 + 30 = 180, the third angle of the triangle is 30°. The exterior angle to this angle forms a linear pair. The exterior angle is therefore 180 − 30 = 150°. If you chose answer **a**, you gave the measure of the third angle, the interior angle. If you chose answer **b**, you incorrectly thought that the angles in a triangle sum to 200°, and gave what you thought was the interior angle. Answer **d** is the sum of the interior angles of a triangle, not the measure of a single interior or exterior angle.

27. b. (Lesson 10) Figure A is a scalene triangle. Scalene triangles have no sides equal, no angles equal, and no lines of symmetry. Figure B is a circle. Circles have an infinite number of lines of symmetry. Any diameter is a line of symmetry. Figure C is a rectangle. A rectangle has lines of symmetry through the midpoints of opposite sides. That is, they can be bisected perpendicularly through the longer sides and through the shorter sides. Figure D is a trapezoid. Only an isosceles trapezoid has a line of symmetry because it has two congruent base angles, and this particular trapezoid is not an isosceles trapezoid. Figure E is an isosceles triangle. Isosceles triangles have a line of symmetry between the midpoint of the base side and the vertex of the vertex angle. Choice **a** is incorrect because Figure A also has no lines of symmetry. Choice **c** is wrong because a circle (Figure B) and a rectangle (Figure C) do have lines of symmetry. Choice **d** is incorrect because an isosceles triangle (Figure E) does have a line of symmetry.

28. c. (Lesson 10) When parallel lines are crossed by a transversal line, there are eight angles formed, as shown in the figure. There are several types of angle pairs that are formed in this figure. Angles $\angle 5$ and $\angle 8$ are vertical angles. Vertical angles are formed when two lines intersect. They share the same vertex and have no other points in common; they are opposite each other. Vertical angles are congruent; that is, they have the same measure. Therefore the measure of $\angle 5$ is 32°, the same measure as $\angle 8$. Angles $\angle 4$ and $\angle 5$ are alternate interior angles. When two parallel lines are intersected by a transversal, the alternate interior angles are congruent, so angle $\angle 4$ also has a measure of 32°. Angle $\angle 2$ is supplementary to $\angle 4$. The two angles sum to 180° because they are a linear pair. They form a straight angle (straight line). Thus, angle $\angle 2 = 180° - 32° = 148°$. Choice **a** is incorrect. As stated before, $\angle 4$ is congruent to $\angle 5$; thus it is also 32°. In addition, angle $\angle 2$ has a measure of 148°. Choice **b** is incorrect. The measure of angles $\angle 5$ and $\angle 4$ = 32°, not 148°, because angles $\angle 8$ and $\angle 5$ are vertical angles. Choice **d** is incorrect. Angle $\angle 4$ is congruent to $\angle 5$; thus the measure is also 32°.

29. c. (Lesson 11) A cube has six faces. The surface area of a cube = $6e^2$, where e = the length of the side. The area of each face is 5 mm × 5 mm = 25 mm^2. Then, 25 mm^2 × 6 = 150 mm^2. Choice **a** is the result of multiplying the length of the side only by 6. Choice **b** is the volume of the cube. Choice **d** is the result of finding the correct surface area, but adding 6 to it.

30. b. (Lesson 11) The formula for the area of a rectangle is $A = b \times h$ (base × height) or $A = l \times w$ (length × width). Thus the area of this rectangle is: $(12 \text{ m})(24 \text{ m}) = 288 \text{ m}^2$. Choices **a** and **d** are incorrect, as they each square one side of the rectangle instead of multiplying length times width. Choice **c** is incorrect—72 is the perimeter of the rectangle, not the area.

31. b. (Lesson 11) To calculate the length of the hypotenuse of a right triangle, use the Pythagorean theorem, which states that the square of the hypotenuse is equal to the sum of the squares of the other two sides of the right triangle. The formula is $a^2 + b^2 = c^2$, where c is the longest side (hypotenuse) of the right triangle and a and b are the other two sides. For this problem:

$9^2 + 12^2 = c^2 = 81 + 144 = 225$

$\sqrt{225} = \sqrt{c^2}$

$15 = c$

Choice **a** is incorrect and may be the result of using the value of 12 as the longest side of the triangle. Choice **c** is the result of simply adding the two known sides together. Choice **d** is incorrect because it neglects to take the square root of 225; $225 = c^2$.

32. a. (Lesson 12) Slope is the ratio of the change in the y-coordinate to the change in the x-coordinate between two points (x_1, y_1) and (x_2, y_2). This is also known as $\frac{rise}{run}$. Substitute in the values of the points to get the slope = $\frac{y_2 - y_1}{x_2 - x_1} = \frac{1 - -2}{3 - 5} = \frac{3}{-2} = -\frac{3}{2}$. The slope of the line is $-\frac{3}{2}$, answer choice **a**. If you chose answer **b**, you added the x-coordinates instead of subtracting. Choice **c** is incorrect because it puts the change in the x-coordinates in the numerator instead of the denominator. If you chose answer **d**, you did the integer arithmetic wrong and got +2 instead of −2.

33. c. (Lesson 12) The distance, d, between two points on the coordinate plane (x_1, y_1) and (x_2, y_2) is $d = \sqrt{(x_2 - x_1)^2 + (y_2 - y_1)^2}$. Substitute the values of the points given into the formula to get:

$d = \sqrt{[(-2) - (-3)]^2 + (3 - 2)^2} = \sqrt{(1)^2 + (1)^2}$
$= \sqrt{1 + 1} = \sqrt{2}$. If you chose answer **a**, you forgot to take the square root. If your answer was choice **b**, you added the coordinates instead of subtracting. If you chose answer **d**, you added the coordinates instead of subtracting, and then also did not square the terms before adding.

34. c. (Lesson 13) The graphical solution to this system of equations is the coordinates of the point of intersection of the two graphed lines. This point is one unit to the left of the origin and five units above the origin. The solution is $(-1, 5)$. If your answer was choice **a**, you incorrectly named the y-coordinate first and the x-coordinate second. If you chose answer **b**, you gave the y-intercept of one of the graphed lines. If your answer was choice **d**, you forgot the negative sign on the x-coordinate.

35. b. (Lesson 13) There is an infinite number of solutions to this system of inequalities: any point in the darkest shaded area. The only point that satisfies this criterion is $(5, 6)$. Choice **a** is on the borderline of the inequalities; this line is dotted, so it is not in the solution set. If your choice was answer **c**, this point is in the solution set of one of the inequalities, $(y > \frac{1}{2}x + \frac{1}{2})$, but not in the intersection of the inequalities. Choice **d** is a point in the solution of $x > 3.5$, but not in the intersection of the two inequalities.

36. a. (Lesson 13) You can use the substitution method of solving the system of equations. Use the first equation and substitute in the given value for y into the second equation:

$$2y \qquad + 4x = 19 \quad \text{Write the original second equation.}$$
$$2(x-4) + 4x = 19 \quad \text{Substitute the first equation in for the value of } y.$$
$$2x - 8 \;\; + 4x = 19 \quad \text{Distribute the 2.}$$
$$6x - 8 \qquad = 19 \quad \text{Combine like terms.}$$
$$\underline{+\;\;\;\; 8 \qquad =+8} \quad \text{Add 8 to both sides.}$$
$$6x \qquad = 27 \quad \text{Simplify.}$$
$$\tfrac{6x}{6} \qquad = \tfrac{27}{6} \quad \text{Divide both sides by 6.}$$
$$x \qquad = 4.5$$

Use the value of 4.5 for x and solve for y in the first equation to get $y = 4.5 - 4 = 0.5$. The solution is (4.5,0.5). Answer choice **b** is a solution to the first equation, but not the second equation, and therefore not a solution to the system. If your answer was choice **c**, you did not distribute the 2 to the 4 when using the distributive property, and got an incorrect value for x. Answer choice **d** is a solution to the first equation but not the system of equations.

37. a. (Lesson 14) First, look for a greatest common factor (GCF) to factor out of each term. The GCF of this polynomial is $3x$. Factoring out $3x$ results in the expression $3x(x^2 + 2x + 1)$. Next, check to see if the expression inside the parentheses can be factored. Use the sum/product rule (in this case, a sum of 2 and a product of 1 give the numbers 1 and 1) to factor the parentheses to $(x + 1)(x + 1)$. Choice **b** has incorrect signs on the terms in parentheses. Choice **c** is not factored. Choice **d** is the greatest common factor only.

38. b. (Lesson 14) The complete product is formed by multiplying each term of the first polynomial by each term of the second polynomial and combining like terms while making sure all positive and negative signs are properly accounted for. Use the distributive property. $(x^2 + 3x - 2)(x^2 + 3x - 2) = (x^2)(x^2) + (x^2)(3x) + (x^2)(-2) + (3x)(x^2) + (3x)(3x) + (3x)(-2) + (-2)(x^2) + (-2)(3x) + (-2)(-2)$ Combine any like terms: $x^4 + 3x^3 - 2x^2 + 3x^3 + 9x^2 - 6x - 2x^2 - 6x + 4 = x^4 + 6x^3 + 5x^2 - 12x + 4$, which is choice **b**. Choice **a** is incorrect: it neglects to add one of the two $3x^3$ terms present in the final problem. Choice **c** is incorrect: it neglects to combine $-6x$ and $-6x$ correctly as $-12x$. Choice **d** is incorrect: it neglects to combine the three x^2 terms correctly: $-2x^2 + 9x^2 - 2x^2 = 5x^2$.

39. d. (Lesson 15) Multiply the numerators, and then the denominators. Cancel any common factors and subtract the exponents when dividing common bases.

$$\left(\tfrac{5x^3}{4}\right)\left(\tfrac{12}{20x}\right) = \left(\tfrac{\cancel{5}^{1}x^3}{\cancel{4}_{1}}\right)\left(\tfrac{\cancel{12}^{3}}{\cancel{20}_{4}\,x}\right) = \tfrac{3x^3}{4x^1} = \tfrac{3x^2}{4}$$

Therefore, the answer is $\tfrac{3x^2}{4}$. Choice **a** is incorrect because it is not fully simplified: $\tfrac{6x^3}{8x} = \tfrac{3x^2}{4}$. Choice **b** is incorrect because it incorrectly simplifies the x factors: $x^3 \div x = x^2$. Choice **c** is incorrect because it incorrectly simplifies the numerical factors: $5 \div 20 = \tfrac{1}{4}$.

40. c. (Lesson 15) First, factor the expressions. $(x^2 + 3x - 10) = (x + 5)(x - 2)$ and $(x^2 - 2x - 15) = (x + 3)(x - 5)$, so the expression becomes $= \frac{(x+5)(x-2)}{x+3} \div \frac{x+5}{(x+3)(x-5)}$. Next, change the division to multiplication and take the reciprocal of the second fraction. Cancel any common factors:

$$\frac{(\cancel{x+5})(x-2)}{\cancel{x+3}} \times \frac{(\cancel{x+3})(x-5)}{\cancel{x+5}} = (x-2)(x-5)$$

$= x^2 - 7x + 10$, which is answer choice **c**. Choice **a** is incorrect and could be the result of improperly multiplying $(x - 2)(x - 5)$. Choice **b** is incorrect because it inverts both factors to multiply, not just the second factor. Choice **d** is incorrect because it does not invert the second factor to multiply.

41. d. (Lesson 16) Substitute the value of 12 for x into the function $f(x) = -3x + 10$, to get $-3(12) + 10 = -36 + 10 = -26$. If your answer was choice **a**, you ignored the variable x and just evaluated $-3 + 10$. If your answer was choice **b**, you did not multiply -3 and 12 correctly, getting $+36$ instead of -36. You would arrive at the incorrect answer of choice **c** if you added -3 and 12 instead of multiplying.

42. c. (Lesson 16) First, substitute the value of -4 into the $g(x)$ function. That results in $g(-4) = -4 + 1 = -3$. Use this value of -3 in the $f(x)$ function, which yields $(-3)3 - 1 = -27 - 1 = -28$. If you chose answer **a**, you incorrectly evaluated the $g(x)$ function and got -5 instead of -3. If your answer was choice **b**, you incorrectly evaluated the $g(x)$ function and got $+3$ instead of -3. If your answer was choice **d**, you did not perform the exponent operation in the $f(x)$ function, but multiplied -3 and 3, instead of $(-3)^3$.

43. a. (Lesson 16) Choice **a** is correct. The value of x is given as -1. Then, the base $\frac{1}{4}$ is being raised to the power of the variable x, or -1. When there is a negative exponent, you take the reciprocal of the base and make the exponent positive. It is then multiplied by -5, and 3 is subtracted from it. Thus, $f(-1) = -5(\frac{1}{4})-1 - 3 = -5(4) - 3 = -20 - 3 = -23$. The ordered pair is $(-1, -23)$. If your answer was **b**, you incorrectly multiplied $-5(4)$ as $+20$. If your answer was choice **c**, you forgot to subtract 3. If you chose answer **d**, you did not take the reciprocal of $\frac{1}{4}$.

44. b. (Lesson 17) Choice **b** is correct because graph B depicts a typical exponential graph that hugs the x-axis before turning sharply up or down to depict growth or decay by a given proportion over a set interval. It also passes through the point $(0,1)$. Choice **a** is incorrect because graph A is a straight line (linear function) with the general formula $f(x) = mx + b$. Choice **c** is incorrect because graph C is a quadratic function (parabola) with the standard formula $f(x) = ax^2 + bx + c$. Choice **d** is incorrect because graph D is a linear function with the general formula $f(x) = mx + b$.

45. c. (Lesson 17) To solve $\log_3 n - \frac{\log_3 1}{4} = \log_3 36$, first use the Laws of Logarithms: $\frac{\log_3 n}{\frac{1}{4}} = \log_3 36$. Divide by $\frac{1}{4}$ to get $\log_3 4n = \log_3 36$. Since both sides are in terms of log base 3, just solve for $4n = 36$. Divide both sides by 4 to get $n = 9$. If your answer was choice **a**, you multiplied 36 by 4 instead of dividing. If you chose answer **b**, you multiplied by 4 and also did not cancel out the log function. If your answer was choice **d**, you did not cancel out the log function.

46. d. (Lesson 17) To solve an exponential equation where the bases are not similar powers, you can use common logarithms. To solve $4^x = 24$ to the nearest hundredth, note that 4 and 24 do not have similar powers. So take the common logarithm of each side and use the Laws of Logarithms:

4^x	$= 24$	Write the original equation.
$\log 4^x$	$= \log 24$	Take the common logarithm of each side.
$x \log 4$	$= \log 24$	Use the power logarithm rule.
$x \frac{\log 4}{\log 4}$	$= \frac{\log 24}{\log 4}$	Divide both sides of the equation by $\log 4$ to isolate x.
$\frac{\log 24}{\log 4}$	$\approx \frac{1.38}{0.60}$	Use the calculator to find the common logarithms.
x	≈ 2.29	Divide and round the answer.

Choice **a** is the result of simply dividing 24 by 4. Choice **b** is also incorrect and could be the result of multiplying 24 by 4. Choice **c** is incorrect and could be obtained by dividing $\log 4$ by $\log 24$, which is the incorrect process.

47. b. (Lesson 18) Choice **b** is correct. Remember SOH–CAH–TOA:

$\sin = \frac{opposite}{hypotenuse}$

$\cos = \frac{adjacent}{hypotenuse}$

$\tan = \frac{opposite}{hypotenuse}$

We are given that $\sin \alpha = \frac{17}{21} = \frac{opposite}{hypotenuse}$ and that $\cos \alpha = \frac{15}{21} = \frac{adjacent}{hypotenuse}$, so we know that $opposite = 17$ and $adjacent = 15$. Therefore, $\tan \alpha = \frac{opposite}{adjacent} = \frac{17}{15}$. Choice **a** is incorrect because $\tan \alpha = \frac{opposite}{adjacent} = \frac{17}{15}$. Choice **c** is incorrect because this is $\frac{hypotenuse}{adjacent}$, not $\frac{opposite}{adjacent}$. Choice **d** is incorrect because this is $\frac{hypotenuse}{opposite}$, not $\frac{opposite}{adjacent}$.

48. d. (Lesson 19) Treat this equation as you would a quadratic equation—write the equation in standard form, and temporarily take out the cosine function: $\cos^2 x + 3\cos x = -2$ becomes $z^2 + 3z + 2 = 0$. Then factor the left side of the equation: $z^2 + 3z + 2 = (z + 2)(z + 1)$. The factors are $(z + 2)(z + 1)$, so replacing the cosine functions into the factored equation results in: $(\cos x + 2)(\cos x + 1) = 0$. Set each factor equal to zero and solve for x using the inverse functions.

$\cos x + 2 = 0$ or $\cos x + 1 = 0$

$\cos x = -2$ or $\cos x = -1$

$\cos^{-1}(-2)$ has no solution or $\cos^{-1}(-1) = 180°$ ($\cos x$ cannot be < -1)

Choice **a** is incorrect because $\cos 0° = 1$, not -1. Choice **b** is incorrect because $\cos 90° = 0$, not -1. Choice **c** is incorrect because it includes the incorrect solution of $0°$.

49. c. (Lesson 20) Choice **c** is correct. First, you must convert the negative radicals into imaginary form:

$$\frac{\sqrt{-32}}{\sqrt{-64}} = \frac{4i\sqrt{2}}{8i}$$

You can cancel the common factor of $4i$ from the numerator and the denominator to get $\frac{\sqrt{2}}{2}$. If your answer was choice **a**, you disregarded the square root operation. If you chose answer **b**, you incorrectly thought that you could just divide the radicands. If your answer was choice **d**, you disregarded the 4 in the numerator.

50. a. (Lesson 20) Choice **a** is correct. Use the function rule to find the first and last elements. $n_1 = -2(1) + 5 = 3$ and $n_{20} = -2(30) + 5 = -35$. Use these values in the formula:

$$\sum_{i=1}^{n} a_i = \left(\tfrac{n}{2}\right)(a_1 + a_n)$$

$\left(\tfrac{20}{2}\right)(3 + -35) = 10(-32) = -320$

If your answer was choice **b**, you just gave the last element in the partial sum sequence. If your answer was choice **c**, you gave the partial sum of the first two elements instead of the first 20 elements. If you chose answer **d**, you may have incorrectly thought you needed to solve an equation, such as $-2n + 5 = 20$.

1 ▶ NUMBER SENSE
For the COMPASS, ASSET, and ACCUPLACER

Everything that can be counted does not necessarily count;
everything that counts cannot necessarily be counted.
—ALBERT EINSTEIN

LESSON SUMMARY

The study of the number system and the sets that make it up is essential to understanding mathematics and how numbers work. This chapter reviews the different types of number sets, as well as the correct order of operations that must be used to calculate accurately. In addition, you will learn about the concepts of factorials, permutations and combinations, and averages (mean, median, and mode).

Number Sets

Understanding the sets of numbers that comprise our number system is a solid way to begin studying for placement exams. The number system is broken down into categories, as shown in the following graphic.

Real Numbers

Rational Numbers   Irrational Numbers

Integers

Whole Numbers

Natural (counting) Numbers

DEFINITIONS

Real numbers: The set of all rational and irrational numbers.

Rational numbers: Any number that can be expressed as $\frac{a}{b}$, where a and b are integers, and $b \neq 0$. Rational numbers include all repeating and terminating decimals, and all fractions, integers, whole numbers, and natural numbers. You will find explanations about and practice with rational numbers in Lesson 3, as well as other lessons in this book.

Irrational numbers: The set of all nonrepeating, nonterminating decimals. The value of π is an irrational number, as is the square root of any nonperfect square, such as $\sqrt{2}$.

Integers: The set of whole numbers and their opposites—these include negative numbers, but do not include fractions or decimals. The set of integers is the primary topic in Lesson 2.

Whole numbers: The set containing the numbers {0, 1, 2, 3, 4, 5, . . .}. These do not include negative numbers.

Natural (counting) numbers: The set containing the numbers {1, 2, 3, 4, 5, . . .}. Note that the only difference between whole numbers and natural numbers is that the natural numbers do not contain the number 0. This is because of the fact that naturally, if we begin to count, we begin with the number 1 and not 0.

Imaginary (complex) numbers: This is the set of numbers whose squares are negative, including $i = \sqrt{-1}$. This particular set of numbers will be explained in detail in Lesson 20.

Two important types of whole numbers are **prime** and **composite**.

DEFINITIONS

Factors are the integers that divide into another integer without a remainder.

Prime numbers are whole numbers whose only factors are 1 and the number itself. Some examples are 2, 3, 5, 7, and 11.

Composite numbers are whole numbers with more than two factors. The number 6 is a composite number because 1, 2, 3, and 6 are all factors of 6.

These types of numbers will be used and incorporated into various lessons in this book and will also be found in questions on many college placement exams.

Order of Operations

When evaluating a mathematical expression, you must perform the operations in a correct sequence. This is called the order of operations.

TIP

The order of operations is commonly remembered with the acronym **PEMDAS**, which stands for:

Parentheses (or any grouping symbols)
Exponents
Multiply/**D**ivide in order from left to right
Add/**S**ubtract in order from left to right

It can also be remembered with the saying "**P**lease **E**xcuse **M**y **D**ear **A**unt **S**ally."

These are just quick, easy ways to remember the correct order, but the examples that follow will explain the process.

Example 1

$4 + 12 \div 3$

Follow the correct order of operations. Since there are no parentheses or exponents, divide 12 by 3 to get 4 first. The problem becomes:

$4 + \underline{12 \div 3}$

$ 4 + \underline{4}$

Now, add 4 + 4 to get the final answer of 8.

Example 2

$7 + (10 - 2) \times 4$

Follow the correct order of operations. Since there is an operation within parentheses, subtract 10 − 2 first. The problem becomes:

$7 + \underline{(10 - 2)} \times 4$

$ 7 + \underline{8} \times 4$

Next, multiply 8 by 4 to get 32. The problem becomes:

$7 + \underline{8 \times 4}$

$ 7 + \underline{32}$

Now, add 32 + 7 to get the final answer of 39.

Example 3

$7^2 - (6 + 2) + 20$

Follow the correct order of operations. First, evaluate the expression within the parentheses. The expression becomes:

$7^2 - \underline{(6 + 2)} + 20$

$ 7^2 - \underline{8} + 20$

Next, evaluate the exponent. 7^2 means $7 \times 7 =$ 49. The expression now becomes:

$\underline{7^2} - 8 + 20$

$49 - 8 + 20$

Now, add and subtract in order from left to right. $49 - 8 = 41$, and then $41 + 20 = 61$. The final answer is 61.

Example 4

$|12 - 10| + 14(3)$

Follow the correct order of operations. In this question, the bars represent the absolute value, or the distance the number is away from 0 on a number line. Because the bars act as a grouping symbol, subtract 12 − 10 = 2 first.

$|12 - 10| + 14(3)$

$|2| + 14(3)$

The absolute value of 2 is 2, and the product of 14 and 3 is 42, so the problem becomes:

$2 + 42 = 44$

Practice

Evaluate each problem using the order of operations.

1. $6 + 2 \times 5$

2. $10 \div 2 - 3$

3. $12 - (3 + 2)$

4. $4^2 + 14 \div 7$

5. $(6 - 3) \times (4 - 1)$

6. $25 - (4 + 1)^2$

7. $8^2 + 3 - |10 \times 2|$

Counting Problems

On your exam, you will likely see problems that deal with the number of ways certain objects can be grouped, arranged, or selected. In each of these, **factorials** are used. Let's define these first.

Factorials

A **factorial** is represented by a whole number followed by an exclamation point. The expression *n*! is defined as $n \times (n-1) \times (n-2) \times \ldots \times 1$. This means to multiply starting with the given number (n) and then to decrease by 1 each time. Always stop the factors when you get to the number 1.

Example

$2! = 2 \times 1 = 2$

$3! = 3 \times 2 \times 1 = 6$

$5! = 5 \times 4 \times 3 \times 2 \times 1 = 120$

QUICK FACT

$0! = 1$

Permutations

Permutations deal with the number of ways objects can be arranged, when the order makes a difference. The number of arrangements can be found by making a list. For example, if you are arranging the letters A, B, and C in all the ways possible, the various orders could be ABC, BCA, CBA, ACB, BAC, and CAB—there are six possible permutations.

Example

If you are arranging four books on a shelf, the number of ways they can be arranged is that there are four choices for the first book, then three choices for the second book, then two choices for the third book, and then only one book left for the final spot on the shelf. This can be summarized as:

$4! = 4 \times 3 \times 2 \times 1 = 24$ ways

This is known as a permutation of four objects taken four at a time, or $_4P_4$.

However, if there are four books to choose from and room for only two of them on the shelf, they can be selected in only $_4P_2 = 4 \times 3 = 12$ ways.

QUICK FACTS

The number of permutations (arrangements) of n objects taken n at a time is $_nP_n = n!$

The number of permutations (arrangements) of n objects taken r at a time is $_nP_r = \frac{n!}{(n-r)!}$. As in the previous example, $_4P_2 = \frac{4!}{2!} = \frac{4 \times 3 \times 2 \times 1}{2 \times 1} = 4 \times 3 = 12$.

Combinations

The number of combinations of a set of objects pertains to the number of groups of the objects, when the order does *not* make a difference.

Example

If three people need to be selected from eight for a committee, then there are $8 \times 7 \times 6 = 336$ ways to select the three people. That is, there are eight possibilities for the first choice, seven for the second, and six for the third. However, the number of the different orders that the three people can be selected needs to be divided out, because the same three people may be chosen in a number of different orders, and the order that they are selected does not matter. This makes the problem $\frac{8 \times 7 \times 6}{3 \times 2 \times 1} = \frac{336}{6} = 56$. There are 56 different ways to select a committee of three people from a total of eight people.

QUICK FACT

The formula for the number of combinations of n objects taken r at a time is $_nC_r = \frac{n!}{r!(n-r)!}$. As in the previous example, $_8C_3 = \frac{8!}{3!5!} = \frac{8 \times 7 \times 6 \times 5 \times 4 \times 3 \times 2 \times 1}{3 \times 2 \times 1 \times 5 \times 4 \times 3 \times 2 \times 1} = \frac{336}{6} = 56$. Remember that the order does not matter in a combination problem.

Practice

Evaluate each of the following.

8. $1!$

9. $6!$

10. $8!$

11. $_3P_3$

12. $_7P_2$

13. $_5P_5$

14. $_4C_2$

15. $_{10}C_8$

16. $_5C_5$

17. How many ways can three pictures be arranged in a row on a wall when selected from a total of five pictures?

18. How many different orders of the digits 1, 2, 3, and 4 can be arranged to make a code of three digits, if no digits repeat?

19. How many combinations of three players can be selected from a team of seven players?

20. How many combinations of two different toppings for a pizza can be selected from a total of nine toppings?

Averages

Mean

The *mean* of a set of numbers is often called the *average*, but is actually just one of several types of averages that give information about a set of numbers. To find the mean of a set of numbers, add the entire set of numbers together and then divide by the total number of elements in the set.

Example 1

To find the mean of the set $\{3, 7, 8, 10\}$, first find the sum of the set of numbers:

$$3 + 7 + 8 + 10 = 28$$

Now, since there are four numbers in the set, divide $28 \div 4 = 7$. The mean of the set of numbers is 7.

Here is another example of a question involving the mean of a set of numbers.

Example 2

If Carla earns the scores 82, 74, and 90 on the first three tests in a class, what does she need to earn on the fourth test to have an average of exactly 85?

The total number of points she would need to earn on four tests to have an average of 85 is equal to $85 \times 4 = 340$ points. One way to solve this problem is to calculate the total number of points she has earned on the first three tests, and then subtract this amount from the average of 85 multiplied by 4. The total number of points she has earned is $82 + 74 + 90 = 246$. The total she needs is 340, so $340 - 246 = 94$. She needs a 94 on the fourth test to have an average of exactly 85.

Median

The median of a set of numbers is another type of average. The median is the middle number in the set when the numbers are arranged in ascending or descending order.

Example

Take the set {4, 8, 5, 6, 7}. First, arrange the numbers in ascending order, or least to greatest:
        4, 5, 6, 7, 8
Next, find the middle number in the set. Since there are five numbers in the set, the third value is the middle number. This number is 6, so 6 is the median of the set of numbers.

When finding the median of a set that contains an even number of elements in it, the process is a bit different. Take the set {11, 12, 14, 16}. This set contains four elements; therefore, there are two values in the middle of the set. To find the median, take those two values and find their mean. In this set, the two numbers in the middle of the set are 12 and 14. The mean of 12 and 14 is:

$$12 + 14 = 26$$
$$26 \div 2 = 13$$

The median of this set of numbers is 13. There are occasions where the median of a set of data is not an element of the set, and this is one of those cases. If the sum of the two middle numbers is odd, the median of a set might even be a decimal.

Mode

Another type of average is the mode. The mode of a set of numbers is the number that occurs the most in the set.

Example

In the set {2, 3, 6, 5, 3, 4, 3} the number 3 is the mode of the set because it appears more than any of the other numbers in the set.

A set can also be bimodal, meaning there are two modes. The set {20, 30, 20, 40, 30} contains the modes 20 and 30, as they both occur twice in the set. This set is bimodal.

If there is no value that appears more times than any other value in the set, then the set is considered to have no mode. For example, the set {1, 2, 3, 4, 7, 11} has no mode since each value appears exactly once and no value appears more than that. Note also that a set may have no mode, even if the numbers are repeated in the set, as long as each number is repeated an equal number of times. For example, {2, 8, 3, 8, 2, 3} has no mode.

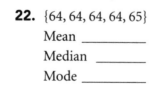

QUICK FACT

The types of numbers known as **averages** (mean, median, and mode) are also known as *measures of central tendency*, or *measures of center*.

Practice

Identify the mean, median, and mode for Questions 21 through 25.

21. {4, 5, 6}
Mean _____
Median _____
Mode _____

22. {64, 64, 64, 64, 65}
Mean _____
Median _____
Mode _____

23. {4, 15, 12, 17, 12, 20}
Mean _____
Median _____
Mode _____

24. {1, 5, 100}

Mean _____

Median _____

Mode _____

25. {8}

Mean _____

Median _____

Mode _____

26. Serina's final average in a course is determined by finding the mean of her four quarterly grades and her final exam, and they are all weighted the same. If she has earned 91, 86, 89, and 95 as her four quarterly grades, what does she need to earn on the final exam to have a final average of exactly 92?

Practice Answers

1. 16

2. 2

3. 7

4. 18

5. 9

6. 0

7. 47

8. 1

9. 720

10. 40,320

11. 6

12. 42

13. 120

14. 6

15. 45

16. 1

17. 60

18. 24

19. 35

20. 36

21. 5, 5, no mode

22. 64.2, 64, 64

23. $13.\overline{3}$, 13.5, 12

24. $35.\overline{3}$, 5, no mode

25. 8, 8, 8

26. 99

2 ▶ INTEGERS AND OPERATIONS
For the COMPASS, ASSET, and ACCUPLACER

It takes these very simple-minded instructions—"Go fetch a number, add it to this number, put the result there, perceive if it's greater than this other number"—but executes them at a rate of, let's say, 1,000,000 per second. At 1,000,000 per second, the results appear to be magic.

—STEVE JOBS

LESSON SUMMARY

Integers are a subset of the real numbers that were defined in Chapter 1. This chapter reviews the concept of integers and guides you to fluency in working with positive and negative numbers. Working with integer operations is straightforward once you have the rules clear in your mind.

We live in a world of checks and balances—it is a world of debits and credits, where quantities are being raised and lowered. Many people have debt—a credit card or a home mortgage, for instance. In addition, they also earn money in the form of income. Temperatures rise and fall, and if you live in the far north, your temperatures are sometimes negative degrees. In order to make sense of your world, an understanding of positive and negative numbers is essential. It is best to hone these skills first with the integers, and then in Chapter 3 you will apply these rules to rational numbers.

Absolute Value

Absolute value is defined as a number's distance from 0 on a number line. Distance is *always* a positive value. Whether your car goes forward for 3 miles or you travel 3 miles in reverse, in either case your odometer will show an increase of 3. The absolute value of a number is represented by two vertical lines around the number, such as |20|.

As shown on the following number line, both 5 and –5 are 5 units from zero. Therefore, |5| = 5 and |–5| = 5.

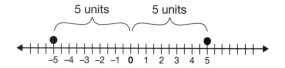

TIP

When you see the symbol "–" before a number or term of an expression, it can be considered as being *minus* or *negative* or the *opposite*. In this way, you can think of –(–6) as the opposite of negative 6, which is 6.

Practice
Simplify:

1. |–27|

2. –|–18|

3. –(–24)

4. What is the opposite of 10?

5. What is the additive inverse of –16?

QUICK FACTS

The absolute value symbol is a pair of vertical lines: |x|. This symbol is a grouping symbol, as described in Chapter 1 when you reviewed the order of operations. Therefore, when you see the absolute value symbol, you perform the operations within the vertical lines first, and then make the resulting expression positive.

Opposite

The opposite of a number is the number that you would add to it to get 0. It is also referred to as the additive inverse. Opposite numbers have the same absolute value, but opposite signs.

For example, the opposite of 7 is –7 and the opposite of –15 is 15.

Ordering Integers

You can easily order integers if you think of them as locations on the number line. Look at the following number line. Negative numbers are to the left of 0 on the number line, and positive numbers are to the right of 0 on the number line. The further to the left a number is, the smaller the number is. Therefore, –16 is smaller than, or less than, –2 because it is to the left of –2 on the number line. But be careful! The absolute value of –16 is greater than the absolute value of –2 because absolute value is always positive, and 16 is greater than 2.

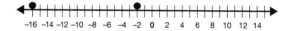

Practice

Insert the correct symbol: <, >, or =.

6. −22 _____ −24

7. |−22| _____ |−24|

8. −|−15| _____ −(−15)

Adding Integers

The rules for adding integers are straightforward, as long as you follow these steps:

> If the numbers have the same sign, *add* their absolute values and keep the sign.
>
> If the numbers have different signs, *subtract* their absolute values (larger minus smaller) and take the sign of the number with the larger absolute value.

You can see this on a number line. An arrow above the number line represents a number. The length of the arrow is the absolute value of the number, and its direction represents the sign of the number.

For example, to add −7 to −6, the numbers have the same sign, so there is no overlap on the arrows. This shows addition of the numbers.

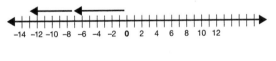

$$-7 + -6 = -13$$

Now consider the problem −7 + 3. The numbers are not the same sign, so there is an overlap on the arrows of the number line, which shows subtraction, or the difference between the arrow lengths. The answer is negative because $|-7| > |3|$.

$$-7 + 3 = -4$$

Subtracting Integers

In order to subtract integers, recall that subtracting is actually adding the opposite. Therefore, to subtract, you change subtraction to addition and change the second term to its opposite. Then simply follow the rules for addition as shown earlier.

For example, to simplify 10 − 16, change the problem to 10 + −16. Because the numbers have different signs, subtract to get −6.

To simplify 25 − −4, change the problem to 25 + 4 = 29.

To simplify −18 − 4, change the problem to −18 + −4 = −22.

Practice

Simplify:

9. −30 + 12

10. −5 + −15

11. 27 + −17

12. 14 − 20

13. $-18 - 5$

14. $12 - -12$

QUICK FACTS

To add or subtract signed numbers:

Is the problem subtraction?

If so, change the problem to addition and make the second number its opposite.

Do the numbers have the same sign?

If so, add and keep the sign.

If not, subtract and take the sign of the number with the larger absolute value.

The rules for multiplication and division are easier to remember than the rules for addition and subtraction.

Multiplication of Integers

To multiply signed numbers, consider the signs of the numbers, just as you did for addition.

If the numbers have the same sign, multiply, and the product is positive.

If the numbers have different signs, multiply, and the product is negative.

For example, $-8 \times -10 = 80$ and $-12 \times 6 = -72$.

You can think of the rule for multiplication in another way:

If there is an even number of negative factors, the answer is positive. Otherwise the answer is negative. (Remember that 0 is even.) This version of the rule is helpful if there are more than two factors, or if there is an exponent involved. We will look at exponents in Chapter 6.

To simplify $-4 \times 3 \times -2 \times 6$, there are two negative factors, -4 and -2. This is an even amount, so multiply the four factors and the answer is positive. Thus, $-4 \times 3 \times -2 \times 6 = 144$.

Another example is to simplify $-4 \times -5 \times -7$. There are three negative factors, which is an odd amount. Multiply and the product is negative. Therefore, $-4 \times -5 \times -7 = -140$.

Dividing Integers

The rule for dividing signed numbers is the same as for multiplying. The sign will be determined according to whether the dividend and divisor have the same or different signs.

Therefore $-180 \div -6 = 30$ because -180 and -6 have the same sign. The quotient is positive. On the other hand, $200 \div -10 = -20$, because the dividend and the divisor have different signs.

REMEMBER

For division: If the numbers have the same sign, the result is positive. Otherwise, the result is negative.

For multiplication, if the numbers are positive or there is an even number of negative factors, the product is positive. Otherwise, the result is negative.

Practice
15. $10 - (-7)(-6)$

16. $-9(-2 + 7)$

17. -35×-2

18. 24×-3

19. $-54 \div -9$

20. $-121 \div 11$

21. $-10 \times -4 \times 3$

22. $-6 \times 3 \times -2 \times -10$

23. $(-5 - 7) \div 4$

24. $-45 + 50 \div -5$

25. $-350 \div (-17 + 12) \times 3$

26. $80 \div -8 \times 5 + |-4 - 2|$

Practice Answers

1. 27
2. −18
3. 24
4. −10
5. 16
6. >
7. <
8. <
9. −18
10. −20
11. 10
12. −6
13. −23
14. 24
15. −32
16. −45
17. 70
18. −72
19. 6
20. −11
21. 120
22. −360
23. −3
24. −55
25. 210
26. −44

L E S S O N

3  RATIONAL NUMBERS
For the COMPASS, ASSET, and ACCUPLACER

A man is like a fraction whose numerator is what he is and whose denominator is what he thinks of himself. The larger the denominator, the smaller the fraction.

—LEO TOLSTOY

LESSON SUMMARY
Fractions and decimals are some of the most frequently used number types, and knowing the basics is very important in the study of more advanced mathematics. This lesson reviews the different types of fractions, and explains how to compare and order fractions and decimals, and perform operations using them. Application problems that involve fractions and decimals are very common on college placement exams.

Fractions

QUICK FACT

Common fractions have two parts. The top number is known as the **numerator**, and the bottom number is known as the **denominator**. The numerator represents the number of pieces a fraction contains, and the denominator indicates the number of equal parts that make up the whole.

There are a variety of different fraction types, including proper fractions, improper fractions, and mixed numbers.

DEFINITIONS

Proper fractions are fractions that can be expressed as $\frac{a}{b}$, where $b \neq 0$, and $a < b$. This just means that the smaller number is in the numerator. Some examples are $\frac{1}{2}$, $\frac{3}{4}$, and $\frac{5}{8}$.

Improper fractions are also expressed in the form $\frac{a}{b}$, where $b \neq 0$; however, in this case, $a \geq b$. In other words, the numerator is greater than or equal to the denominator. For example, the fractions $\frac{9}{8}$, $\frac{20}{17}$, and $\frac{7}{5}$ are each improper fractions.

Mixed numbers contain an integer with a proper fraction following. A few examples are $4\frac{2}{3}$, $-3\frac{7}{8}$, and $12\frac{1}{7}$.

Converting Mixed Numbers and Improper Fractions

To convert a mixed number into improper form, multiply the integer part by the denominator, and then add the numerator. This value becomes the numerator of the fraction, and the denominator remains the same.

Example 1

The mixed number $3\frac{5}{7} = \frac{3 \times 7 + 5}{7} = \frac{21 + 5}{7} = \frac{26}{7}$.

To convert from an improper fraction to a mixed number, divide the numerator by the denominator. The quotient becomes the integer part of the mixed number and the remainder becomes the numerator of the fractional part.

Example 2

In the fraction $\frac{20}{3}$, divide 20 by 3 and get 6 with a remainder of 2. This converts the improper fraction to $\frac{20}{3} = 6\frac{2}{3}$. Sometimes it may be necessary to simplify the fraction part when performing this conversion.

Practice

Convert to an improper fraction.

1. $2\frac{1}{2}$

2. $4\frac{3}{5}$

3. $8\frac{2}{7}$

Convert to a mixed number.

4. $\frac{32}{15}$

5. $\frac{23}{7}$

6. $\frac{100}{99}$

Simplified Fractions

Simplified fractions have a numerator and denominator with no factors in common, and they can be proper or improper. Some examples are $\frac{1}{4}$, $\frac{5}{3}$, and $\frac{12}{17}$.

When the greatest common factor of two distinct numbers is 1, the numbers are called *relatively prime*. Simplified fractions have a numerator and denominator that are relatively prime.

9. $3\frac{20}{30}$

10. $-\frac{14}{22}$

11. $\frac{-8}{-24}$

Fractions that are not simplified have a common factor that needs to be divided out.

Example

In the case of the fraction $\frac{10}{12}$, the numerator and denominator have a common factor of 2. Divide each by this factor to simplify the fraction to $\frac{10 \div 2}{12 \div 2} = \frac{5}{6}$.

QUICK FACTS

The **greatest common factor (GCF)** of two or more numbers is the largest number that divides evenly into the numbers. The GCF is commonly used to simplify fractions.

The **least common multiple (LCM)** of two or more numbers is the smallest number that the values divide into without a remainder. The LCM is commonly used to get a common denominator (called the *least common denominator*) when comparing fractions and when adding and subtracting fractions.

Practice

Write each of the following in simplest form:

7. $\frac{3}{6}$

8. $\frac{2}{10}$

Comparing Fractions

There are a few ways to compare fractions. When comparing fractions with a common denominator, the larger the numerator, the larger the fraction. For example, $\frac{7}{9} > \frac{4}{9}$. When fractions have different denominators, it is often helpful to convert them to equivalent fractions with a common denominator.

Example 1

Take the two fractions $\frac{2}{3}$ and $\frac{4}{5}$. First, get the least common denominator of 15. This converts the fractions to $\frac{2 \times 5}{3 \times 5} = \frac{10}{15}$ and $\frac{4 \times 3}{5 \times 3} = \frac{12}{15}$. Since 10 is less than 12, then $\frac{10}{15} < \frac{12}{15}$, and therefore $\frac{2}{3} < \frac{4}{5}$.

Another way to compare is to use the cross products. Recall that the word *product* indicates multiplication.

Example 2

For example, compare the fractions $\frac{3}{4}$ and $\frac{5}{8}$. Multiply the numerator of the first fraction by the denominator of the second fraction to get the first cross product ($3 \times 8 = 24$). Then multiply the numerator of the second fraction by the denominator of the first fraction ($5 \times 4 = 20$). Then, compare the products. The first product of 24 corresponds with the first fraction and is the larger of the two numbers, so the first fraction is larger. Thus, $\frac{3}{4} > \frac{5}{8}$.

Practice
Compare each of the following by placing the correct symbol on the blank: <, >, or =.

12. $\frac{2}{5}$ _____ $\frac{3}{5}$

13. $\frac{3}{7}$ _____ $\frac{3}{8}$

14. $\frac{4}{10}$ _____ $\frac{2}{5}$

15. $1\frac{2}{3}$ _____ $1\frac{3}{5}$

16. $-2\frac{1}{5}$ _____ $-2\frac{1}{3}$

Operations with Fractions

Adding and Subtracting Fractions
In order to add and subtract fractions, they need to have a common denominator. Then, the numerators are added or subtracted according to the problem at hand.

Examples
In the following example, there is already a common denominator. Add the numerators and keep the common denominator: $\frac{3}{5} + \frac{1}{5} = \frac{4}{5}$.

In the case of $2\frac{2}{5} - 1\frac{1}{10}$, first get a common denominator of 10.

The problem then becomes $2\frac{4}{10} - 1\frac{1}{10}$.

Subtract the numerators of the fractions, and then the integer parts, but keep the common denominator to get $2\frac{4}{10} - 1\frac{1}{10} = 1\frac{3}{10}$.

When dealing with positive and negative fractions, follow the rules for adding and subtracting integers.

Examples
In the problem $-\frac{2}{3} + -\frac{4}{3}$, the signs are the same, so add and keep the sign. Because there is a common denominator of 3, combine the

numerators and keep the denominator. The answer is $-\frac{6}{3} = -2$.

In the problem $-\frac{1}{2} + \frac{3}{4}$, the signs are different. Convert the fractions to a common denominator of 4, and then subtract the values of the numerators. The problem becomes $-\frac{2}{4} + \frac{3}{4} = \frac{1}{4}$.

Multiplying Fractions
To multiply fractions, simply multiply the numerators together to find the numerator of the product, and multiply the denominators together to find the denominator. Then, simplify the result if necessary.

Example
For example, to multiply $\frac{5}{8} \times \frac{7}{9}$, multiply the numerators of 5 and 7 to get a product of 35, and multiply the denominators of 8 and 9 to get a product of 72. The result becomes $\frac{5}{8} \times \frac{7}{9} = \frac{35}{72}$. Because 35 and 72 are relatively prime, the solution is in simplest form.

TIP
If there are common factors between the numerators and denominators, these can be factored out and simplified before multiplication takes place. In this way, the answer will already be in simplest form. For example, in the following problem the common factor of 3 is divided out, and then the remaining numbers are multiplied across the numerators and denominators: $\frac{\cancel{3}^1}{4} \times \frac{1}{\cancel{9}_3} = \frac{1}{12}$.

Dividing Fractions
To divide fractions, first be sure that all fractions are in proper or improper form, and not in mixed number form. Then, multiply by the reciprocal of the divisor in order to divide. Recall that the reciprocal of a fraction is when the numerator and denominator are switched.

Example 1

In the example $\frac{7}{8} \div \frac{3}{4}$, change the operation to multiplication and take the reciprocal of the fraction following the operation sign. Then complete the problem by multiplying across numerators and denominators:

$$\frac{7}{8} \div \frac{3}{4} = \frac{7}{8} \times \frac{4}{3} = \frac{28}{24} = \frac{7}{6} = 1\frac{1}{6}$$

TIP

Before multiplying or dividing fractions, be sure to change any mixed numbers to improper fractions.

Recall that if there is an even number of negatives in a multiplication or division problem, the answer is positive. If there is an odd number of negatives, the result is negative.

Example 2

The problem $-\frac{1}{6} \times -\frac{1}{5} = \frac{1}{30}$ yields a positive result because there are two negatives. The problem $\frac{3}{5} \div -\frac{1}{5} = \frac{3}{5} \times -\frac{5^1}{1} = -3$ because there is an odd number of negatives.

Practice

Perform the indicated operation.

17. $\frac{2}{3} + \frac{4}{3}$

18. $\frac{12}{13} + 1\frac{1}{26}$

19. $-\frac{14}{15} - \frac{8}{15}$

20. $3\frac{1}{4} - 1\frac{1}{8}$

21. $\frac{20}{27} \times \frac{9}{10}$

22. $6\frac{1}{7} \times -\frac{5}{8}$

23. $\frac{3}{4} \div \frac{5}{6}$

24. $-2\frac{2}{3} \div -3\frac{1}{4}$

Decimals

Decimals are based on place value, as shown in the following diagram.

T H O U S A N D S	H U N D R E D S	T E N S	O N E S	• D E C I M A L P O I N T	T E N T H S	H U N D R E D T H S	T H O U S A N D T H S

To read decimals, use the decimal place furthest to the right. For example, the decimal 0.12 is read as "twelve-hundredths" because it has a value two places to the right of the decimal point. In the same manner, the decimal 3.4 is read as "three and four-tenths." Note that the word *and* is in the place of the decimal point.

Rounding Decimals

To round decimals, look to the place value directly to the right of the decimal place to be rounded to.

Examples

If rounding the decimal 4.362 to the nearest hundredth, look one decimal place to the right. This is the thousandths place, the 2. If the value in this place is 0 to 4, keep the value the same in the hundredths place and drop the numbers to the right of it. If the value in the thousandths place is 5 to 9, increase the value in the hundredths place by one and drop the digits to the right. In the decimal 4.362, 6 is in the hundredths place and 2 is in the thousandths place.

Since 2 is between 0 and 4, keep the 6 the same and drop the digits to the right of the 6. This decimal rounds to 4.36, the nearest hundredth.

As another example, the decimal 5.8713 rounded to the nearest tenth is 5.9. The 7, which is in the hundredths place, is between 5 and 9, so change 8 to 9 and drop the digits to the right of the 9.

Comparing Decimals

In order to compare decimals, use place value. For example, $0.3 > 0.2$ since three-tenths is greater than two-tenths.

The decimal $0.005 < 0.05$ because five-thousandths is less than five-hundredths. The 5 in 0.05 has a larger place value than the 5 in 0.005.

The decimal $1.3 > 1.26$ because 1.3 is equivalent to $1\frac{30}{100}$ whereas 1.26 is equivalent to $1\frac{26}{100}$.

Practice

25. Round the decimal 54.89 to the nearest tenth.

26. Round the decimal 9.016 to the nearest hundredth.

Compare each of the following by using <, >, or =.

27. 0.4 _____ 0.7

28. 0.123 _____ 0.12

29. 3.2 _____ 3.18

Adding and Subtracting Decimals

In order to add or subtract decimals, the most important thing to remember is to line up the decimal points.

Example

To add 0.23 + 3.4, first line up the decimal points vertically. A 0 can be placed after the number 3.4 to convert it to 3.40 to help line up the decimal places. Then, perform the operation. The problem becomes:

$$\begin{array}{r} 0.23 \\ + \underline{3.40} \\ 3.63 \end{array}$$

As with fractions, follow the rules for adding and subtracting integers when adding or subtracting decimals. For example, $-0.2 + -0.3 = -0.5$. Since the signs are the same, add the decimals and keep the sign. In the problem $-1.54 + 4.54$, subtract the decimals and take the sign of the decimal with the larger absolute value. The result is 3.

TIP

When adding or subtracting decimals, always be sure that the decimal points are lined up vertically. Add 0's as placeholders to help line up the decimal places.

Multiplying and Dividing Decimals

When multiplying decimals, multiply the values as you normally would if there were no decimal points. Then, count the number of digits to the right of the decimal point in each number being multiplied. Move the decimal this total number of places in the product.

Example

In the problem 3.5×2.1, multiply the values as usual.

$$\begin{array}{r} 3.5 \\ \times \underline{2.1} \\ 3\,5 \\ \underline{70\,0} \\ 73\,5 \end{array}$$

Now, place the decimal point between the 7 and the 3 in the product since there was a total of two decimal places to the right of the decimal points in the numbers being multiplied. The final answer is 7.35. A good way to check that you've placed the decimal point correctly is by rounding: 3.5 is close to 4 and 2.1 is close to 2. You know that $4 \times 2 = 8$, so the answer should be in the ballpark of 8. If you got an answer of 73.5, then your decimal point is in the wrong place.

DEFINITIONS

The **divisor** is the number you divide by in a division problem.

The **dividend** is the number being divided into in a division problem.

The **quotient** is the result/answer to a division problem.

For example, in the problem $72 \div 8 = 9$, 72 is the dividend, 8 is the divisor, and 9 is the quotient.

When dividing decimals, first move the decimal point in the divisor so that the divisor is a whole number. In addition, move the decimal point in the dividend the same number of places to the right. Then, place the decimal directly above its location in the dividend. This will be the location of the decimal in the final answer, or quotient.

Example

Take the division problem $4.53 \div 0.3$. First, set up the division problem and change the divisor to 03. Place the decimal point to the right one place in the dividend, and then place a decimal point directly above it where the quotient will be. Complete the problem using division.

$$0.3\overline{)4.53} \text{ becomes } 03.\overline{)4\,5.3}$$

$$\begin{array}{r} 15.1 \\ 03.\overline{)4\,5.3} \\ \underline{-3} \\ 15 \\ \underline{-15} \\ 0\,3 \\ \underline{-3} \\ 0 \end{array}$$

TIP

When multiplying a number by a power of 10, just move the decimal point to the right the same number of places as zeros. For example, multiplying by 10 moves the decimal point one place to the right, and multiplying by 100 moves the decimal point two places to the right. When dividing a number by a power of 10, just move the decimal point to the left in the same fashion.

Recall that if there is an even number of negatives, the result is positive. If there is an odd number of negatives, the result is negative. For example, $-0.24 \div -0.6 = 0.4$ and $1.2 \times -6 = -7.2$.

Practice

Perform the indicated operation.

30. $6.8 + -3.73$

31. $9.12 - 4.51$

32. 3.2×-7.5

33. $-12.1 \div -1.1$

34. 2.34×10

35. 7.213×100

36. $56.34 \div 10$

37. $120.9 \div 100$

Changing Decimals to Fractions

Repeating and terminating decimals have an equivalent fraction, based on the place value of the decimal. For example:

$$0.3 = \frac{3}{10} \text{ (three-tenths)}$$
$$0.43 = \frac{43}{100} \text{ (43 hundredths)}$$
$$0.291 = \frac{291}{1,000} \text{ (291 thousandths)}$$

Changing Fractions to Decimals

One way to convert a fraction to a decimal is to convert the fraction to a decimal-place denominator such as 10 or 100.

Example

Take the fraction $\frac{2}{5}$. Convert this fraction to a denominator of 10 by multiplying both numerator and denominator by 2. The fraction becomes $\frac{2 \times 2}{5 \times 2} = \frac{4}{10}$. Because this fraction is equal to 4 out of 10, it is equivalent to four-tenths, or 0.4.

TIP

The bar in any fraction means division, so another way to change a fraction to a decimal is to divide the numerator by the denominator. In the fraction $\frac{1}{4}$, divide 1 by 4 to get 0.25, the decimal form of the fraction.

```
   0.25
4)1.00
  -8
  ──
   20
  -20
  ───
    0
```

In an example such as $\frac{2}{3} = 2 \div 3 = 0.66666666$ . . . , place a bar over the digit or digits that repeat: $\frac{2}{3} = 0.\overline{6}$.

Comparing Fractions and Decimals

When comparing a fraction and a decimal, first decide on a common form of the numbers (fraction or decimal) and relate them to each other.

Example 1

To compare 0.24 and $\frac{1}{5}$, the fraction $\frac{1}{5}$ can be converted to decimal form. Divide 1 by 5 to get 0.2. Now, compare the two values. 0.24 > 0.2, so $0.24 > \frac{1}{5}$.

Another way to compare decimals and fractions is to convert both to fraction form.

Example 2

To compare $\frac{11}{20}$ and 0.57, first change the denominator of the fraction to 100. The fraction $\frac{11}{20}$ becomes $\frac{11 \times 5}{20 \times 5} = \frac{55}{100}$. Convert the decimal 0.57 to the fraction $\frac{57}{100}$. Because $\frac{55}{100} < \frac{57}{100}$, then $\frac{11}{20} < 0.57$.

TIP

When comparing rational numbers by changing to fraction form, choose denominators of 10, 100, 1,000, and so on when possible.

Practice

Compare each of the following by using <, >, or =.

38. $\frac{33}{50}$ _____ 0.66

39. 0.29 _____ $\frac{8}{31}$

40. 3.41 _____ $\frac{11}{3}$

41. $-\frac{2}{3}$ _____ −0.67

42. −11.3 _____ $-11\frac{1}{3}$

Applications

Solve each of the following problems. Express each answer in simplest form.

43. What is $\frac{1}{3}$ of 78?

44. When purchasing two shirts for $8.95 each, how much change from a $20 bill will you get back?

45. Order the following from least to greatest:
$0.4, \frac{1}{4}, 0.24, \frac{11}{25}$

46. $\left(\frac{2}{5} + \frac{1}{3}\right) - \frac{1}{15}$

47. Sharon purchased $\frac{1}{2}$ lb. of Swiss cheese, $1\frac{1}{4}$ lb. of American cheese, and 0.75 lb. of cheddar cheese. In pounds, what is total amount of cheese she has purchased?

Practice Answers

1. $\frac{5}{2}$
2. $\frac{23}{5}$
3. $\frac{58}{7}$
4. $2\frac{2}{15}$
5. $3\frac{2}{7}$
6. $1\frac{1}{99}$
7. $\frac{1}{2}$
8. $\frac{1}{5}$
9. $3\frac{2}{3}$
10. $-\frac{7}{11}$
11. $\frac{1}{3}$
12. $<$
13. $>$
14. $=$
15. $>$
16. $>$
17. 2
18. $1\frac{25}{26}$
19. $-\frac{22}{15} = -1\frac{7}{15}$
20. $2\frac{1}{8}$
21. $\frac{2}{3}$
22. $-3\frac{47}{56}$
23. $\frac{9}{10}$
24. $\frac{32}{39}$
25. 54.9
26. 9.02
27. $<$
28. $>$
29. $>$
30. 3.07
31. 4.61
32. -24
33. 11
34. 23.4
35. 721.3
36. 5.634
37. 1.209

38. =

39. >

40. <

41. >

42. >

43. 26

44. $2.10

45. $0.24, \frac{1}{4}, 0.4, \frac{11}{25}$

46. $\frac{2}{3}$

47. $2\frac{1}{2}$ lbs. or 2.5 lbs.

4 ▶ RATIO AND PROPORTION
For the COMPASS, ASSET, and ACCUPLACER

Every man is rich or poor according to the proportion between his desires and his enjoyments.

—SAMUEL JOHNSON

LESSON SUMMARY

In our world, we are constantly comparing things. We compare the number of teachers to the number of students in a classroom. We compare the proportion of nuts to raisins in a granola bar, or the number of males to females in a given occupation. In this lesson, we review the concepts of ratio and proportion. We look at a special ratio called a rate (another special ratio called a percent is covered in Lesson 5). Finally, we study common applications of proportions, including conversions, similar figures, and scale.

Ratio

A ratio is a way to compare things with numbers—it represents a relationship between two quantities. There are three different ways to show a ratio: with a colon between the quantities, with the word *to* between the quantities, or as a fraction. For example, if there are eight female nurses for every 10 total nurses, then the ratio of female nurses to total nurses is 8:10, or 8 to 10, or $\frac{8}{10}$.

QUICK FACTS

When working with ratios, be sure you understand what you are comparing. For example, if there are six male dentists for every 10 total dentists, you can have several different ratios, or relationships.

- $\frac{6}{10}$ is the ratio of male dentists to total dentists.
- $\frac{4}{10}$ is the ratio of female dentists to total dentists.
- $\frac{4}{6}$ is the ratio of female dentists to male dentists.

TIP

Because a ratio can be expressed as a fraction, you can simplify a ratio, or show equivalent ratios, just as you do for fractions. In the preceding example, the ratio of female dentists to male dentists can be simplified to $\frac{2}{3}$.

Practice

There are three different colors of jellybeans in a huge vat at a factory. There are 25 red jellybeans to every 100 jellybeans. There are 30 green jellybeans for every 100 jellybeans. The rest are purple. Find the following ratios:

1. Red to green jellybeans

2. Red to the total amount

3. Red to purple

4. Purple to green

5. Purple to the total amount

Rate

A rate is a special ratio that compares different units. You deal with rates often in your everyday life. There are many examples of rates when dealing with automobiles. Speed, such as miles per hour, is a rate. When you go to purchase a car, you are often concerned with miles per gallon, or revolutions per minute. You find many different rates in the grocery store. Many items sell for dollars per pound, or items per dollar.

A unit rate indicates how much of one quantity is related to one unit of the second quantity. So a unit rate has a 1 in the denominator of the fraction. The denominator is often not written when it is 1. All of the examples in the previous paragraph are unit rates. It is easy to compare rates when they are converted to unit rates.

Examples

If five pounds of chicken cost $17.00, then it is selling for $17.00 for five pounds, or $\frac{17.00}{5}$, which is a unit rate of $3.40 per pound.

If you drive 385 miles and it takes you seven hours, then your average rate, also known as speed, is 385 miles per seven hours, or 385 ÷ 7, a unit rate of 55 miles per hour.

Practice

6. A truck travels 432 miles in six hours. What is the average speed of the truck, as a unit rate?

7. Shopper's Heaven sells avocados that are three for $5.00. What is the unit price for an avocado?

8. Beth drives her car a total distance of 1,541 miles in 23 hours of driving time. Which vehicle was traveling faster, Beth's car or the truck in Question 6?

Proportion

A proportion is a mathematical equation that states that two ratios are equal. Just as a ratio compares two quantities, proportions compare two ratios. If two ratios are equivalent, then they are said to be in proportion. The two quantities that make up the ratios have the same relationship to one another. Proportions are generally written as two equal fractions:

$$\frac{w}{x} = \frac{y}{z}$$

Multiply both of these fractions by the common denominator of xz: $xz \times \frac{w}{x}$ and $\frac{y}{z} \times xz$. By canceling the common factors, you find that $zw = yx$. You may recall this as the common procedure known as cross multiplication, where the product of the means, yx, is equal to the product of the extremes, zw.

Proportions are used to solve many types of problems. You can use cross multiplication to test whether two ratios are equivalent, that is, whether they are in proportion.

Example 1
For example, if there are 30 raisins to every 20 cashews in a trail mix, you can set up a proportion to check if a larger bag with 1,950 raisins and 1,300 cashews is the same ratio: $\frac{30}{20} = \frac{1,950}{1,300}$. Cross multiply to test: $30 \times 1,300 = 20 \times 1,950$, and $39,000 = 39,000$, so the ratios of raisins to cashews are equal.

TIP

When setting up a proportion, be sure to keep like quantities in the same relative position. Using the previous example, the ratios were both set up as $\frac{raisins}{cashews}$.

You can also find missing values if you know that ratios are proportional.

Example 2
For example, if a car has driven 170 miles in 2.5 hours, how long will it take the car to travel 374 miles at the same rate? You are told that the car is traveling at the same rate, the same ratio. Set up a proportion with miles in the numerator and hours in the denominator: $\frac{170}{2.5} = \frac{374}{x}$. Cross multiply to get $170x = 374 \times 2.5$, or $170x = 935$. Since 170 times the unknown hours is equal to 935, divide 935 by 170 to determine the missing hours: $935 \div 170 = 5.5$ hours.

TIP

When working with rates and units, be sure the units are consistent.

Example 3
If 4 ounces of sesame seeds are sold for $1.20, then to determine how much 2 pounds of sesame seeds cost, you can convert 4 ounces to 0.25 pound: $\frac{1.20}{0.25} = \frac{x}{2}$. Cross multiply to get $1.20 \times 2 = 0.25x$, or $2.40 = 0.25x$. Divide 2.40 by 0.25 to get $9.60.

Practice
9. Navel oranges cost $2.99 for four. At that rate, how much will 18 oranges cost, to the nearest penny?

10. A car travels 112.5 miles in two and a half hours. At this speed, how long will it take to travel 76.5 miles?

11. If you walk at the steady rate of three miles every 45 minutes, how far can you walk in three hours?

12. A recipe for peanut butter cookies calls for 1.5 cups of brown sugar to make 72 cookies. How much brown sugar is needed to make 252 cookies?

Applications

There are many different types of problems that can be solved by proportions.

Example

Consider this type of problem: Two numbers are in the ratio of 3:5, and their sum is 112. What are the numbers? It is given that the sum is 112, and the parts are related to the whole as $\frac{3}{8}$ and $\frac{5}{8}$. You can set up a proportion to find one of the numbers. To find the smaller number, solve: $\frac{3}{8} = \frac{x}{112}$. Cross multiply to get $3 \times 112 = 8x$, or $336 = 8x$. Divide 336 by 8 to get the smaller number, which is 42. The larger number is $112 - 42 = 70$.

Scale

Scale is a special type of ratio that compares a model to a real object. The model could be a map, a scale drawing, a toy, or an architectural model.

Example

For example, a toy car is made to the scale of 1:64. This means that for every 1 unit of the toy, the corresponding dimension of the real car is 64 times as big. For example, if a toy truck is 1.5 inches, you can use a proportion to determine the real length of the truck by setting up the proportion: $\frac{1}{64} = \frac{1.5}{x}$. Cross multiply to get x, the length of the real truck: $x = 64 \times 1.5 = 96$ inches, or divide by 12 inches to get 8 feet long.

TIP

When computing a scale factor, be sure that the quantities being compared are in the same units.

Practice

13. On a map, the scale is 1 cm = 20 miles. If the distance on the map from one city to another is 17 cm, how far apart are the actual cities in miles?

14. A souvenir statue of the Eiffel Tower is 7 inches tall. If the real Eiffel Tower is 1,050 feet in height, what is the scale factor?

15. The scale on a blueprint is 1:360. If the length of a building is 165 feet, what is the length in inches of the building on the blueprint?

Unit Conversions

You can convert units using proportions and cross multiplication. By using this procedure, it is easily apparent whether you should multiply or divide. A conversion factor is a special type of scale factor.

Example

For example, given that 5,280 feet is equivalent to 1 mile, you can convert 110,880 feet to miles by setting up the proportion: $\frac{5,280}{1} = \frac{110,880}{x}$. Cross multiply to get $110,880 = 5,280x$. Divide 110,880 by 5,280 to get 21 miles.

Practice

16. The conversion rate for the United States dollar to the euro is 1 euro = 1.32 dollars. Jennifer is traveling to France and takes $540. How much money is this in euros?

17. There are 5,280 feet in a mile. There are three feet in a yard. How many yards are there in 12.4 miles?

18. There are 1,000 grams in a kilogram. Susan weighs 58,400 grams. How much is her weight in kilograms? There are approximately 2.2 pounds in a kilogram. How much is her weight in pounds?

Similar Figures

Similar figures are figures that have the same shape but a different size. Because of this fact, similar figures have sides that are in proportion. You can test if two figures are similar by testing if the sides are in proportion.

Example 1

For example, to determine if the two rectangles shown here are similar, find the corresponding sides. Set up a proportion by deciding which dimensions should go in the numerator and which in the denominator.

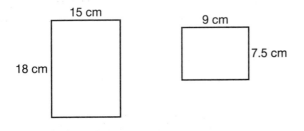

In this example, the shorter side is placed in the numerator and the longer side is placed in the denominator: $\frac{15}{18} = \frac{7.5}{9}$. You may notice that $\frac{15}{18} \div \frac{2}{2} = \frac{7.5}{9}$. This demonstrates that the ratios are equivalent. Or, you can demonstrate that the figures are in proportion by cross multiplying: $9 \times 15 = 7.5 \times 18$, or $135 = 135$.

If you know that figures are similar, you can solve for a missing side measure.

Example 2

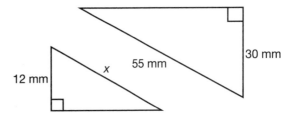

To find the missing side, x, of these similar triangles, set up a proportion: $\frac{12}{x} = \frac{30}{55}$. Cross multiply to get $30x = 660$. Divide 660 by 30 and $x = 22$ mm.

A common word problem using similar triangles is a shadow problem. An object and its shadow form a right angle. By connecting the end of the shadow with the end of the object, you have formed a triangle. Here is a figure of a building and a flagpole. Each casts a shadow and the similar triangles are formed.

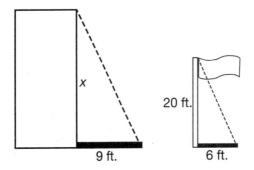

You are given the height of the flagpole the lengths of the shadows. You can determine the height of the building from this information: $\frac{6}{20} = \frac{9}{x}$. Cross multiply and $6x = 180$, and $x = 30$ ft.

Practice

19. Given the dimensions of the following triangles, are the triangles similar?

 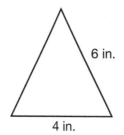

5 in.
3 in.
6 in.
4 in.

20. The trapezoids shown here are similar. Find the missing side, x.

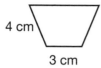 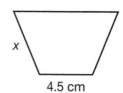

4 cm
3 cm
x
4.5 cm

21. Find the missing side, y, of the parallelogram, to the nearest tenth of a millimeter. The figures are similar.

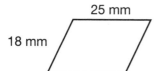 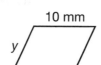

25 mm
18 mm
10 mm
y

22. A tree casts a shadow that is 3.6 feet long. At the same time of day, a 6 foot, 3 inch tall man casts a shadow that is 1.2 feet long. How tall is the tree?

Practice Answers

1. $\frac{25}{30} = \frac{5}{6}$

2. $\frac{25}{100} = \frac{1}{4}$

3. $\frac{25}{45} = \frac{5}{9}$

4. $\frac{45}{30} = \frac{3}{2}$

5. $\frac{45}{100} = \frac{9}{20}$

6. 72 miles per hour

7. $1.67 per avocado

8. Beth was driving 67 miles per hour. The truck was driving faster.

9. $13.46

10. 1.7 hours

11. 12 miles

12. 5.25 cups

13. 340 miles

14. 1,050 feet = 12,600 inches, so the scale factor is $\frac{7}{12,600} = \frac{1}{1,800}$, or 1:1,800.

15. 165 feet = 1,980 inches. 1,980 ÷ 360 = 5.5 inches.

16. 409.09 euros

17. 21,824 yards

18. 58.4 kilograms, or 128.48 pounds

19. No, they are not similar, because $\frac{3}{5} \neq \frac{4}{6}$.

20. Solve $\frac{3}{4} = \frac{4.5}{x}$ to get x = 6.

21. Solve $\frac{18}{25} = \frac{y}{10}$ to get y = 7.2.

22. Draw a picture:

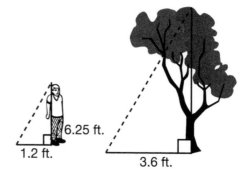

6.25 ft.
1.2 ft.
3.6 ft.

Solve $\frac{1.2}{6.25} = \frac{3.6}{x}$ to get the height of the tree, which is 18.75 ft.

PERCENT AND APPLICATIONS
For the COMPASS and ACCUPLACER

Genius is one percent inspiration, ninety-nine percent perspiration.

—THOMAS EDISON

LESSON SUMMARY

Percents, an important part of mathematics, are found in numerous places in everyday life. Things such as discounts and sale prices, statistics, commissions, bank rates, taxes, and tips all involve percents. As implied by the sum of the values in the Edison quote, the number 100 is critical when studying percent. This lesson reviews the concept of percent, along with conversions between percents, fractions, and decimals. You will also learn applications of percent, such as percent of change, discounts, sale prices, taxes, and gratuities.

Percent

A **percent** is a special ratio that compares a part to a whole, but the whole is always the same.

DEFINITION

A **percent** is a value compared to 100.

The symbol used for percent is "%," and is always written after the value. For example, some common percents are 10%, 25%, 50%, and 100%.

Changing Percents to Decimals

In order to change a percent to a decimal, divide the value of the percent by 100.

For example, $60\% = \frac{60}{100} = 60 \div 100 = 0.60$. Recall from Lesson 3 that dividing by 100 moves the decimal point two places to the left.

TIP

To change a percent to a decimal, simply move the decimal point two places to the left in the percent and remove the % sign. For example, 62% becomes 0.62, 150% becomes 1.5, and 7% becomes 0.07.

Changing Decimals to Percents

When changing from a decimal to a percent, multiply the value of the decimal by 100. For example, $0.45 = 0.45 \times 100 = 45\%$. It was also summarized in Lesson 3 that multiplying by 100 moves the decimal point two places to the right.

TIP

To change a decimal to a percent, move the decimal point two places to the right and place the percent symbol to the right of the number. For example, 0.54 becomes 54%, 0.08 becomes 8%, and 0.341 becomes 34.1%.

Practice

Change each percent to a decimal.

1. 12%

2. 5%

3. 236%

4. 100%

Change each decimal to a percent.

5. 0.79

6. 0.01

7. 2.35

8. 0.002

Changing Percents to Fractions

To change percents to fractions, simply place the value of the percent in the numerator of the fraction with a denominator of 100.

Example
For example, $25\% = \frac{25}{100}$. Then simplify if necessary. In this case, divide both numerator and denominator by the greatest common factor of

25. $\frac{25 \div 25}{100 \div 25} = \frac{1}{4}$. In the case of the percent 62.5%, place 62.5 over 100 and simplify the fraction. $\frac{62.5}{100} = \frac{625}{1,000} = \frac{625 \div 125}{1,000 \div 125} = \frac{5}{8}$

Changing Fractions to Percents

One way to change a fraction to a percent when the denominator is a factor of 100 is to convert the fraction to an equivalent fraction with a denominator of 100. Then, the numerator becomes the percent.

Example
For example, for the fraction $\frac{2}{5}$, convert 5 to a denominator of 100 by multiplying the numerator and denominator by 20. The fraction becomes $\frac{2 \times 20}{5 \times 20} = \frac{40}{100}$. Since the value of 40 is being compared to 100, the percent is 40%.

Another way to convert a fraction to a percent is to change the fraction to a decimal first, and then move the decimal point two places to the right as explained previously. For example, the fraction $\frac{3}{4}$ is equivalent to the decimal 0.75 after dividing $3 \div 4$. Then, move the decimal point two places to the right to get the percent 75%.

QUICK FACT

Fractions that are mixed numbers or improper fractions are equivalent to percents that are greater than or equal to 100%.

Practice
Change each percent to a fraction in simplest form.

9. 10%

10. $33.\overline{3}\%$

11. 128%

12. 500%

Change each fraction to a percent.

13. $\frac{1}{2}$

14. $\frac{3}{8}$

15. $\frac{12}{10}$

16. $3\frac{1}{4}$

Finding the Percent of a Number

There are a few different ways to find the percent of a number. One way is to change the percent to a decimal and multiply. Recall that the key word *of* indicates multiplication.

Example
Find 25% of 12.

First, change the percent to a decimal and the word *of* to *multiplied by*.

The problem becomes 0.25×12. Multiply to get the answer of 3. So 3 is 25% of 12.

Another way to find the percent of a number is to set up a part/whole proportion, as explained in Lesson 4. In this case, the percent is always placed over the 100.

TIP

The proportion $\frac{part}{whole} = \frac{\%}{100}$ can be used for percent problem solving.

Example

Use the proportion in the preceding Tip box to solve the following problem.

Find 8% of 24.

Set up the proportion using 24 as the whole and placing the 8% over the 100: $\frac{x}{24} = \frac{8}{100}$. Find the cross products to get $100x = 192$. Divide each side of the equation by 100 to get 1.92.

TIP

Use some mental math strategies to help solve problems with certain percents. For instance, to find 10% of a number, calculate that 10% $= \frac{10}{100} = \frac{1}{10}$, so just move the decimal point one place to the left, as you are dividing by 10. This amount can now be doubled to find 20%, tripled to find 30%, and so on. In the case of finding 20% of 50, first find 10%, which is 5. Then, double this amount to find 20%: 20% of 50 is 10.

Finding the Whole When the Percent Is Given

The same part-to-whole proportion can be used to find the whole when the percent is given.

Example

The number 30 is 75% of what number?

In this problem, the part is 30, the percent is 75%, and the whole is unknown.

Set up the proportion $\frac{30}{x} = \frac{75}{100}$. Find the cross products to get $3,000 = 75x$. Divide each side of the equation by 75 to get $x = 40$. Thus, 30 is 75% of 40.

Finding the Percent When the Part and Whole Are Given

In a similar fashion, use the part-to-whole proportion to find the percent when the part and whole are given.

Example

The number 24 is what percent of 96?

In this problem, the part is 24, the whole is 96, and the percent is unknown.

Set up the proportion $\frac{24}{96} = \frac{x}{100}$. Find the cross products to get $2,400 = 96x$. Divide each side of the equation by 96 to get $x = 25$. Thus, 24 is 25% of 96.

Practice

For each problem, find the indicated value.

17. What is 10% of 85?

18. What is 30% of 70?

19. What is 16% of 20?

20. What is 150% of 48?

21. 12 is 80% of what number?

22. 68 is 20% of what number?

23. 40 is what percent of 120?

24. 55 is what percent of 44?

Applications

Percent of Change

Percent of change can be an increase or a decrease. An increase could represent a markup in the price of a product, an increase in population of a city, or the growth of a plant. A percent decrease could represent a discount on a price, a weight loss, or a drop in the stock market.

TIP

The proportion used for percent of change is $\frac{change\ (difference)}{original\ amount} = \frac{\%}{100}$.

The following are two examples of percent of change. The first deals with percent of increase, and the second with percent of decrease. The change is always considered a positive value, even if it is a decrease.

Example 1

The average cost of a gallon of gasoline went from $3.95 to $4.25 during a one-month period. What is the percent of increase of the cost of a gallon of gasoline?

To solve this problem, use the proportion $\frac{change\ (difference)}{original\ amount} = \frac{\%}{100}$. To begin the problem, first subtract to find the difference in the two prices: $4.25 – $3.95 = $0.30. Then set up the proportion: $\frac{0.30}{3.95} = \frac{x}{100}$. Find the cross products to get $30 = 3.95x$. Divide each side of the equation by 3.95 to get $7.595 \approx 7.6\%$. The price increased about 7.6%.

Example 2

The number of players on a team decreased from 25 to 20 players. What is the percent of decrease in the number of players on the team?

To solve this problem, use the proportion $\frac{change\ (difference)}{original\ amount} = \frac{\%}{100}$. To begin the problem, first subtract to find the difference in the number of players: $25 – 20 = 5$. Then set up the proportion: $\frac{5}{25} = \frac{x}{100}$. Find the cross products to get $500 = 25x$. Divide each side of the equation by 25 to get 20%. The number of players decreased by 20%.

Tax

Finding the amount of tax that needs to be paid is the same as finding the percent of a number.

Example

If the tax is 7%, what is the total cost of a bike with a price of $250?

To solve this problem, first calculate the amount of tax by finding 7% of $250. Use the proportion $\frac{part}{whole} = \frac{\%}{100}$. Substitute the values to get $\frac{x}{250} = \frac{7}{100}$. Find the cross products to get $100x = 1,750$. Divide both sides of the equation by 100 to get $x = 17.50$. The amount of tax is $17.50. To complete the problem, add the tax to the cost of the bike to find the total cost: $250 + $17.50 = $267.50.

TIP

For some application problems, use estimation to make the problem easier when an exact answer is not necessary. When estimating, the numbers being used are rounded first to more compatible values so the calculation can be done more efficiently.

Gratuity

The term *gratuity* is just another word for a tip that would be left for a service, such as for a server of food, a hairdresser, or a cab driver. Calculating the amount of gratuity that would be given for a service is the same process as finding the percent of a number.

Example

When dining out, a family had a total bill of $62.28. What is the amount of gratuity the family should pay in order to give a 20% tip to the wait person?

In this question, the exact of amount of the gratuity can vary, so use estimation to make the problem easier. First, round the total bill of $62.28 to $60. Next, find 20% of 60 by setting up the proportion $\frac{x}{60} = \frac{20}{100}$ and finding the cross products. The equation becomes $100x =$ 1,200. Divide each side of the equation by 100 to get $x = 12$. The family should leave a gratuity of $12.

TIP

Recall that in this instance 10% of 60 is equal to 6, so 20% of 60 would be twice that amount, or 12. This would be an alternative way to use estimation to quickly find 20% of the amount.

Discounts and Sale Prices

Finding the amount of a discount is the same as finding the percent of the number. To find the sale price, subtract the amount of discount from the original price.

Example

A tablet has a regular price of $349.99. If a store if offering a discount of 25%, what is the sale price of the tablet?

First, find 25% of 349.99. Set up the proportion $\frac{x}{349.99} = \frac{25}{100}$. Find the cross products to get $100x = 8,749.75$. Divide each side of the equation by 100 to get $x = 87.4975$. Recall that this amount can also be found by changing the percent to a decimal and multiplying: $0.25 \times 349.99 = 87.4975$. Round this value to the nearest hundredth to get 87.50. Rounding of decimals was explained in Lesson 3. The total amount of discount is $87.50. Next, subtract this amount from the original price: $349.99 – $87.50 = $262.49. The sale price is $262.49.

Practice

Find the indicated value for each of these problems.

25. An infant's weight was 10.5 pounds at one month and 15.3 pounds a few months later. What is the percent of increase of the infant's weight to the nearest tenth?

26. A city that had a population of 45,606 decreased in population to 42,890. What was the percent of decrease in population of this city, to the nearest tenth?

27. Sierra bought three items that cost $4.60, $12.95, and $10.50. If the tax rate is 6%, what is the total tax she paid on all three items?

28. Kevin has a total bill of $50 when his pet is groomed. If he leaves 15% as a tip, how much gratuity should he give?

29. Kathleen purchases a new jacket with a discount of 20%. If the regular price of the jacket is $75, how much did she save if there is no sales tax?

30. Joan buys a pair of shoes with a regular price of $68 that are on sale with a discount of 30%. What is the sale price of her shoes?

Practice Answers

1. 0.12
2. 0.05
3. 2.36
4. 1
5. 79%
6. 1%
7. 235%
8. 0.2%
9. $\frac{1}{10}$
10. $\frac{1}{3}$
11. $1\frac{7}{25}$
12. 5
13. 50%
14. 37.5%
15. 120%
16. 325%
17. 8.5
18. 21
19. 3.2
20. 72
21. 15
22. 340
23. $33.\overline{3}\% = 33\frac{1}{3}\%$
24. 125%
25. 45.7%
26. 6.0%
27. $1.68
28. $7.50
29. $15.00
30. $47.60

EXPONENTS AND SQUARE ROOTS
For the COMPASS, ASSET, and ACCUPLACER

Anyone who believes exponential growth can go on forever in a finite world is either a madman or an economist.
—KENNETH BOULDING

This chapter covers a lot of ground, and might take you longer than 20 minutes to complete. Split up your study into three parts—spend 20 minutes on the Exponents and Laws of Exponents sections, then 20 minutes on the Negative Exponents and Zero Exponents section and the Scientific Notation section, and finally, 20 minutes on the remainder of the lesson.

LESSON SUMMARY
It is often said that technology and Earth's population are growing *exponentially*. An exponent is a shorthand way of showing repeated multiplication of the same factor. This lesson reviews the laws that govern the use of exponents, as well as negative and fractional exponents. Scientific notation, a common use of exponents, is also covered. Finally, you will learn about square roots and related operations, including simplifying radicals and radical operations.

Exponents

As stated before, an exponent is a shorthand way of showing repeated multiplication of the same factor. For example, $4 \times 4 \times 4 \times 4 \times 4 = 4^5 = 1,024$. We say that "four to the fifth power is 1,024." In the statement $4^5 = 1,024$ the number 4 is called the *base*, the repeated factor. The number 5 is called the *exponent*, or the number of times the base is multiplied. The number 1,024 is the *power*; that is, 1,024 is the fifth power of 4. The base can be a number, a variable, or a variable expression.

Recall from Lesson 2 that the rules for multiplication of integers state that if there is an even number of negative factors, the answer is positive. Otherwise the answer is negative. So if the exponent is an even number, the answer will be positive whether the base number is positive or negative. If the base number is negative and the exponent is an odd number, the answer will be negative.

QUICK FACTS

• An exponent of 1 is usually not shown. For example, x is the same as x^1.
• An expression to the second power has a special name; it is referred to as *squared*. For example, $8^2 = 64$ is read as "eight squared is 64."
• Perfect squares are the whole numbers that result from squaring other whole numbers. Some examples: 1 (1^2), 4 (2^2), 9 (3^2), 16 (4^2), 25 (5^2), and 36 (6^2).
• Likewise, an expression to the third power is referred to as *cubed*, so $2^3 = 8$, which is read as "two cubed is eight."

In Lesson 1, you reviewed the order of operations. Recall that the order of operations is:

Parentheses
Exponents

Multiplication and Division, left to right
Addition and Subtraction, left to right

Be careful to understand that exponentiation is performed before any multiplication.

Example
To find the value of $5m^2$ when $m = 3$, you first find $3^2 = 9$ and then multiply $5 \times 9 = 45$. This is different from the expression $(5m)^2$ when $m = 3$. In this latter case, you first multiply $5 \times 3 = 15$ because parentheses are evaluated first; you then square it to get $15^2 = 15 \times 15 = 225$.

Practice
Simplify:

1. 4^3

2. $(-3)^4$

3. $(-2)^5$

4. 5^2

5. -1^3

Evaluate when $x = 4$:

6. $5x^2$

7. $(5x)^2$

8. $(x - 12) - 5 + x^3$

Laws of Exponents

There are certain laws, sometimes referred to as rules, which govern the use of exponents.

Multiplying Powers with the Same Base

The first law deals with multiplying powers of the same base. Look at the following example.

Example 1

$(3 \times 3) \times (3 \times 3 \times 3 \times 3) = 3^6$. Notice that there are two factors of 3 multiplied by four more factors of 3. This is a total of $2 + 4 = 6$ factors of 3, or 3 to the sixth power. Hence, $3^2 \times 3^4 = 3^{2+4} = 3^6$.

Example 2

Here is another example: $y^3 \times y^2 = (y \times y \times y) \times (y \times y) = y^5$. This is a total of three factors of y multiplied by two more factors of y, which is $3 + 2 = 5$ factors of y, or y to the fifth power. In other words, $y^3 \times y^2 = y^{3+2} = y^5$.

This leads to the first law of exponents: **When you multiply two expressions with the same base, you keep the base and add the exponents.** This law applies only when the expressions have the same base. For example, $x^2 \times y^5$ cannot be simplified. You cannot add the exponents if the bases are different.

Dividing Powers with the Same Base

The second law deals with dividing powers with the same base. Let's look at an example. $\frac{7^6}{7^2} = \frac{7 \times 7 \times 7 \times 7 \times 7 \times 7}{7 \times 7}$. If you simplify this fraction by canceling the common factors in the numerator and the denominator, you are left with $7 \times 7 \times 7 \times 7 = 7^4$. This process of canceling common factors in this way is really subtracting out common factors. So $\frac{7^6}{7^2} = 7^{6-2} = 7^4$. This also is true if the base is a variable: $\frac{n^3}{n} = \frac{n \times n \times n}{n} = n^{3-1} = n^2$. The second law of exponents: **When you divide powers with the same base, keep the base and subtract the exponents.** This law applies only when the expressions have the same base. For example, $\frac{x^7}{y^2}$ is in its simplest form. You cannot subtract the exponents if the bases are different.

Examples

To multiply $\left(\frac{8x^4}{3}\right)\left(\frac{6y}{x^2}\right)$, multiply the numerators and the denominators, and use the laws of exponents: $\frac{(8 \times 6)x^4 y}{3x^2} = 16x^{4-2}y = 16x^2 y$.

To divide $\frac{10m^8}{7n} \div \frac{2m^5}{n^4}$, change the division to multiplication by the reciprocal, and then simplify:

$\frac{10m^8}{7n} \times \frac{n^4}{2m^5} = \frac{10m^{8-5} n^{4-1}}{14} = \frac{5m^3 n^3}{7}$

Practice

Simplify:

9. $2^2 \times 2^4$

10. $(-3) \times (-3)^4$

11. $(-4)^2 \times 5^2$

12. $m^4 \times m^6$

13. $\frac{10^6}{10^3}$

14. $\frac{2^2 \times 2^4}{4^2}$

15. $\frac{x^{10}}{x^4(x)}$

16. $a^4 \times b^6$

17. $a^2 b^3 \times a^4$

18. $\left(\frac{32r^5}{5t^3}\right)\left(\frac{5t^4}{4r^7}\right)$

19. $\frac{6x^5 y}{25} \div \frac{2x^5}{5y^2}$

The Power of a Power, and the Power of a Product

Consider the following expression: $(x^2)^3$. This means to take the base of x^2 and multiply it three times: $x^2 \times x^2 \times x^2$. We know from the first law as reviewed earlier that $x^2 \times x^2 \times x^2 = x^{2+2+2} = x^6$. This is three groups of two, or in other words $3 \times 2 = 6$. So the next law is:

When you take the power of a power, keep the base and multiply the exponents.

The power of a product refers to when the base itself is an expression that is a product. For example, let's look at the expression $(2x^4)^3$. Recall from earlier in this lesson that if you know a value for x, you would evaluate all in parentheses first. If you are not given a value for x, then this law will direct you to apply the power of 3 to all factors in the parentheses. So $(2x^4)^3 = 2^3 \times (x^4)^3 = 8 \times x^{4 \times 3} = 8x^{12}$. Another example is: $-(3m^6n)^2 = -1 \times 3^2 \times m^{6 \times 2} \times n^{1 \times 2} = -9m^{12}n^2$. This law is stated as: **When you take the power of a product, the power applies to each factor of the product.**

Practice

Simplify each:

20. $(2^2)^4$

21. $[(-2)^3]^2$

22. $(x5)^7$

23. $(xy)^4$

24. $(4x)^3$

25. $(10x^2y)^4$

QUICK FACTS

Here are the laws of exponents, expressed with algebra:

$x^r \times x^s = x^{r+s}$

$\dfrac{x^r}{x^s} = x^{r-s}$

$(x^r)^s = x^{rs}$

$(xy)^r = x^r y^r$

Negative Exponents and Zero Exponents

Zero exponents and negative exponents are probably the most difficult aspects of exponents to remember. They can best be understood by viewing what happens when there are positive exponent values in a fraction.

Example

Consider $\frac{5^4}{5^6}$. By the division law of exponents, this fraction is equal to $5^{4-6} = 5^{-2}$. To understand the meaning, expand the factors in the fraction: $\frac{5^4}{5^6} = \frac{5 \times 5 \times 5 \times 5}{5 \times 5 \times 5 \times 5 \times 5 \times 5}$. When you divide out all of the common factors of 5, you are left with $\frac{1}{5 \times 5} = \frac{1}{5^2}$. This leads to the definition of negative exponents: **If there is a negative exponent, take the reciprocal of the base and make the exponent positive.** As you saw earlier, $5^{-2} = \frac{1}{5^2}$. If the base is a whole number, the numerator of the simplified version is 1. If the base is a fraction, the simplified fraction is the reciprocal. So $(\frac{3}{4})^{-2} = (\frac{4}{3})^2 = \frac{16}{9}$.

Recall that any nonzero number divided by itself is equal to 1. So therefore $\frac{6^3}{6^3} = 6^{3-3} = 6^0 = 1$. The definition of a zero exponent is: **Any nonzero number to the zero exponent is 1.** This also holds true for any expression: $(3x^4)^0 = 1$, no matter what the value of x is, providing it is not zero.

Practice

Simplify. If the base is a variable, express the answer with positive exponents.

26. 2^{-5}

27. $(xy)^{-8}$

28. $m^5 \times m^{-9}$

29. $\frac{3^5}{3^8}$

30. $17^6 \times 17^{-6}$

31. $\left(\frac{4}{5}\right)^{-2}$

32. $(2x^{-3}y^2)^4$

Evaluate when $x = -3$ and $y = 4$:

33. $\frac{x^2}{x^5}$

34. $(x^5y^6)^0$

Scientific Notation

Scientific notation is a special, but very specific, way of naming numbers. It is used by scientists and engineers, who deal with very large or very tiny numbers. For example, the approximate number of red blood cells that the typical human has is 25,000,000,000,000. A biologist does not want to constantly rewrite such a number with so many zeros. The diameter of a hydrogen atom is 0.000000000025 meters. Again, a scientist does not want to rewrite all of these zeros. Scientific notation is a convention of naming numbers that is based on the powers of 10. This is convenient, because our decimal number system is based on the powers of 10.

For example, the number of red blood cells, 25,000,000,000,000, is equivalent to 2.5×10^{13}, and the diameter of a hydrogen atom, 0.000000000025 meters, is equivalent to 2.5×10^{-11}. Notice that very large numbers have a positive exponent and very small numbers have a negative exponent. Because the special notation is based on the powers of 10, you can use this knowledge to help you to determine how the exponent in the notation is related to the standard form of the number.

Converting a Number from Scientific Notation to Standard Form

To convert a number from scientific notation to standard form, you simply move the decimal point in the given number the amount of places as specified by the exponent. If the exponent is positive, move the decimal point to the right. If the exponent is negative, move the decimal point to the left.

Examples

For example, to convert 4.03×10^8 to a number in standard form, move the decimal point in 4.03 eight places to the right, to make a large number.

$$4.03000000 = 403{,}000{,}000$$

| Decimal point moved eight places right |

To convert 7.92×10^{-6} to a number in standard form, move the decimal point in 7.92 six places to the left, to make a tiny number.

$$000007.92 = 0.00000792$$

| Decimal point moved six places left |

Practice

Convert to standard form:

35. 7.5×10^7

36. 6.09×10^9

37. 2.4×10^{-5}

38. 1.05×10^{-8}

Converting a Number from Standard Form to Scientific Notation

To convert a number from standard form to scientific notation, again you move the decimal point. You move the decimal point to the place where the number will have a value that is greater than or equal to 1 but less than 10. The number of places you have to move the decimal point determines the power of 10. If the original number was tiny, then the exponent is negative; if the original number was large, then the exponent is positive.

Examples

For example, to convert 502,000,000,000,000 to a number in scientific notation, you must move the decimal point from the end (the far right) of the number to make the number be 5.02, which is between 1 and 10. The decimal moves 14 places, so the number is 5.02×10^{14}. The exponent is positive, because the number was very large.

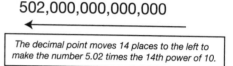

| The decimal point moves 14 places to the left to make the number 5.02 times the 14th power of 10. |

To convert 0.0000425 to scientific notation, you must move the decimal point from where it is at present to make the number 4.25, which is between 1 and 10. The decimal moves five places, so the number is 4.25×10^{-5}. The exponent is negative, because the number is tiny.

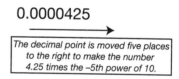

| The decimal point is moved five places to the right to make the number 4.25 times the −5th power of 10. |

Practice

Convert to scientific notation:

39. 7,080,000,000

40. 43,000,000,000,000

41. 0.000000209

42. 0.00056

Practice

Insert the correct symbol: <, >, or =.

43. 4.6×10^{12} _____ 5.2×10^9

44. 3.02×10^{-2} _____ 6.1×10^{-4}

45. 7.32×10^{11} _____ 8.1×10^{11}

Radicals

For every defined operation in mathematics, there is an inverse, or opposite, operation. The inverse of addition is subtraction; the inverse of multiplication is division. The inverse operation to exponentiation is the root, or the radical. The square root is the most common radical; it is the opposite of squaring, or taking a number or expression to the second power.

Examples

For example, because $3^2 = 9$, it is said that the square root of 9, or $\sqrt{9}$, equals 3. The symbol $\sqrt{}$ is called the radical symbol. The number or expression underneath the symbol is called the radicand.

Another example is that because $2^5 = 32$, then $\sqrt[5]{32} = 2$. The small number 5 in the radical symbol is called the index, or the nth root, in this case the fifth root of 32. Because the square root is so commonly used, the index of 2 is not written; it is understood.

Practice

Simplify:

46. $\sqrt{121}$

47. $\sqrt[4]{16}$

48. $\sqrt[3]{125}$

Radical Operations—Square Roots

You can perform the operations of adding, subtracting, multiplying, and dividing with radicals. The expression $5\sqrt{2}$ means five times the square root of 2. You can add two radicals if they have the same radicand. For example, $5\sqrt{2} + 7\sqrt{2} = 12\sqrt{2}$. On the other hand, $\sqrt{2}$ and $\sqrt{3}$ have different radicands, so you cannot add them. Therefore, $7\sqrt{2} + 3\sqrt{5}$ is considered to be in simplified form. Subtraction works in a similar way, so $4\sqrt{7} - 6\sqrt{7} = -2\sqrt{7}$. When you multiply or divide radicals, you can multiply or divide the radicands. For example, $\sqrt{12} \times \sqrt{3} = \sqrt{12 \times 3} = \sqrt{36} = 6$. Similarly, $\dfrac{\sqrt{54}}{\sqrt{6}} = \sqrt{\dfrac{54}{6}} = \sqrt{9} = 3$.

Practice
Simplify:

49. $8\sqrt{3} + 6\sqrt{3}$

50. $2\sqrt{2} - 5\sqrt{2}$

51. $-4\sqrt{11} + 3\sqrt{2} + 7\sqrt{11}$

52. $\sqrt{27} \times \sqrt{3}$

53. $\sqrt{2} \times \sqrt{3}$

54. $\dfrac{\sqrt{80}}{\sqrt{5}}$

Simplifying Radicals—Square Roots

In Lesson 1, rational numbers were defined. The only radicals that are rational numbers are those with radicands that are perfect squares. Most radicals are irrational numbers. Frequently in mathematics radicals are not approximated, but are just simplified. In order to simplify a radical you factor out any and all perfect squares. Just like you can multiply two radicals, you can factor radicals.

Example 1
For example, to simplify $\sqrt{108}$ you can find the prime factorization by making a factor tree:

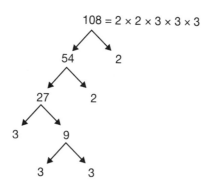

$$108 = 2 \times 2 \times 3 \times 3 \times 3$$

The prime factorization for 108 is $2^2 \times 3^3$. Pair up prime factors to get $2^2 \times 3^2 \times 3$. Because squaring and square root are opposite

operations, $\sqrt{2^2} = 2$, and $\sqrt{3^2} = 3$. There is one prime factor left, a 3. So $\sqrt{108} = \sqrt{4 \times 9 \times 3} = \sqrt{4} \times \sqrt{9} \times \sqrt{3} = 2 \times 3 \times \sqrt{3} = 6\sqrt{3}$.

If you are asked to simplify sums of radicals, you should simplify each radical term. Any of the terms that have the same simplified radical are combined.

Example 2
For example, to add $\sqrt{80} + \sqrt{18} + \sqrt{125}$, simplify each term.

$$\sqrt{80} = \sqrt{16 \times 5} = 4\sqrt{5}$$
$$\sqrt{18} = \sqrt{9 \times 2} = 3\sqrt{2}$$
$$\sqrt{125} = \sqrt{25 \times 5} = 5\sqrt{5}$$

So the sum is $4\sqrt{5} + 3\sqrt{2} + 5\sqrt{5} = 3\sqrt{2} + 9\sqrt{5}$. This is simplified because the remaining radicands are different.

Practice
Simplify:

55. $\sqrt{98}$

56. $\sqrt{243}$

57. $\sqrt{48} + \sqrt{12}$

58. $\sqrt{72} - \sqrt{200}$

59. $\sqrt{50} + \sqrt{28} - \sqrt{32}$

Cube Roots, nth Roots, and Fractional Exponents

The nth root has been reviewed previously. When n is 2, it is called a square root. When n is 3, it is a called a cube root. Roots are often expressed with the radical sign. Another way to express roots is by fractional exponents.

DEFINITIONS

$$x^{\frac{1}{2}} = \sqrt{x}$$
$$x^{\frac{1}{3}} = \sqrt[3]{x}$$
$$x^{\frac{1}{n}} = \sqrt[n]{x}$$

Practice Answers

1. 64

2. 81

3. −32

4. 25

5. −1

6. 80

7. 400

8. 51

9. $2^6 = 64$

10. $(-3)^5 = -243$

11. 400

12. m^{10}

13. 1,000

14. 4

15. x^5

16. a^4b^6

17. a^6b^3

18. $\frac{8t}{r^2}$

19. $\frac{3}{5y}$

20. $2^8 = 256$

21. $(-2)^6 = 64$

22. x^{35}

23. x^4y^4

24. $64x^3$

25. $10,000x^3y^4$

26. $\frac{1}{32}$

27. $\frac{1}{x^8y^8}$

28. $\frac{1}{m^4}$

29. $\frac{1}{27}$

30. 1

31. $\frac{25}{16}$

32. $\frac{16y^8}{x^{12}}$

33. $\frac{1}{x^3} = -\frac{1}{27}$

34. 1

35. 75,000,000

36. 6,090,000,000

37. 0.000024

The definition of a fractional exponent is consistent with the laws of exponents. Recall that squaring a number and taking the square root of a number are opposite operations. One will undo the other.

Example 1

For example, $\sqrt{x^2} = x$. The same example illustrated with fractional exponents is $(x^{\frac{1}{2}})^2 = x^{\frac{1}{2}\times2} = x^1 = x$. Therefore $64^{\frac{1}{2}} = 8$ and $27^{\frac{1}{3}} = 3$.

The numerator of a fractional exponent does not have to be 1.

Example 2

For example, by the fact that $\sqrt[3]{64} = 4$ and $64 = 8^2$, you can substitute to show that $\sqrt[3]{8^2} = 4$ and $\sqrt[3]{8^2} = (8^2)^{\frac{1}{3}}$. By the "power of a power" property, $(8^2)^{\frac{1}{3}} = 8^{\frac{2}{3}}$, so $8^{\frac{2}{3}} = 4$.

Practice

Simplify:

60. $(144)^{\frac{1}{2}}$

61. $(16)^{\frac{1}{4}}$

62. $(\frac{1}{25})^{\frac{1}{2}}$

63. $(49)^{-\frac{1}{2}}$

64. $(16)^{\frac{3}{4}}$

65. $(8)^{-\frac{2}{3}}$

38. 0.0000000105

39. 7.08×10^9

40. 4.3×10^{13}

41. 2.09×10^{-7}

42. 5.6×10^{-4}

43. >

44. >

45. <

46. 11

47. 2

48. 5

49. $14\sqrt{3}$

50. $-3\sqrt{2}$

51. $3\sqrt{2} + 3\sqrt{11}$

52. 9

53. $\sqrt{6}$

54. 4

55. $7\sqrt{2}$

56. $9\sqrt{3}$

57. $4\sqrt{3} + 2\sqrt{3} = 6\sqrt{3}$

58. $6\sqrt{2} - 10\sqrt{2} = -4\sqrt{2}$

59. $5\sqrt{2} + 2\sqrt{7} - 4\sqrt{2} = \sqrt{2} + 2\sqrt{7}$

60. 12

61. 2

62. $\frac{1}{5}$

63. $\frac{1}{7}$

64. 8

65. $\frac{1}{4}$

7 ▶ ALGEBRAIC EXPRESSIONS
For the COMPASS, ASSET, and ACCUPLACER

The human mind has never invented a labor-saving machine equal to algebra.

—Author unknown

LESSON SUMMARY
Learning how to manipulate and simplify algebraic expressions is a basic skill of algebra. This lesson serves as an extension of the previous unit dealing with exponents and reviews elements common to algebraic expressions, such as combining like terms and substituting values for the variable. In addition, you will explore how to translate situations into algebraic expressions and practice how to use formulas.

s the study of algebraic concepts begins, first let's identify the basic parts of any algebraic expression.

DEFINITION

An *algebraic expression* is an expression that contains numbers and/or variables and at least one operation.

QUICK FACT

If no coefficient is written in front of a variable, it is understood to be 1. In the same way, if no exponent is written, it also understood to be 1: $x = 1x^1$.

An algebraic expression can be made up of one or more *terms*. Terms in algebra are separated by addition or subtraction. For example, the expression $3x^2$ has one term, the expression $4z - 5$ has two terms, and the expression $6y^3 + 4y - 2$ has three terms.

QUICK FACT

A simplified algebraic expression with one term is called a *monomial*, with two terms it is called a *binomial*, and with three terms it is called a *trinomial*.

DEFINITION

Algebraic expressions that are *like terms* have exactly the same variable(s) with exactly the same exponent(s).

These expressions are all contained in the more general category of polynomials, which will be studied in greater detail in both this lesson and Lesson 14.

In any monomial, the number in front of the letter is known as the *coefficient*, the letter is known as the *variable*, and the small number raised to the right of the variable or numeral is the *exponent*. The coefficient and variable(s) are connected by multiplication. Exponents were discussed in depth in the previous lesson.

For example, in the expression $9x^3$, the 9 is the coefficient, the x is the variable, and the exponent is 3. Note that a monomial does not have to contain all three of these, such as the monomials $4x$, 7, and x^4.

One of the most common ways to simplify algebraic expressions is to combine like terms.

Here are a few pairs of like terms:

6 and 11
$5x$ and $-7x$
x^2y and $8x^2y$
$12abc$ and $-2abc$

Practice

Determine whether each of the following pairs of terms are like terms. Write *yes* or *no*.

1. $3x$ and $8x$

2. $5a$ and $5b$

3. yz and zy

4. 128 and $-128d$

5. $0.5t^4$ and $1.3t^3$

Simplifying Expressions Using Like Terms

As long as two terms are *like terms*, then they can be combined. This means they can be added or subtracted, depending on the signs. Refer to Lesson 2 for the rules for adding and subtracting with integers for help with positive and negative values.

When combining like terms, the coefficients are added or subtracted and the exponent parts remain the same.

Example 1

$5x + 6x + x$

All of these are like terms because they each have a variable x. Recall that the term x has a coefficient of 1. Add the coefficients and keep the variable part the same:

$$5x + 6x + 1x = 12x$$

Example 2

$10ab^2 - 6ab^2 + 8ab^2$

All of these are like terms since they each have the same variable part, ab^2. Combine the coefficients and keep the variable part the same:

$10 - 6 + 8 = 12$, so the simplified expression is $12ab^2$.

Example 3

$7xy - 4yx + 3$

Even though the variables are in a different order, the first two terms are like terms and can be combined. You can rewrite $4yx$ as $4xy$ because multiplication is commutative. The final term of 3 cannot be combined with the first two terms because it does not have the same variable part, so it continues to be added at the end of the expression. Combine the coefficients of the first two terms and keep the variable part the same:

$7 - 4 = 3$, so the simplified expression is $3xy + 3$.

Example 3

$5s + 5t + 5u$

Because the variable part is not the same for any of the terms, no terms can be combined. The simplified form of this expression is the same, $5s + 5t + 5u$.

Practice

Simplify each of the following by combining like terms.

6. $10x + 11x$

7. $-6d - 2d$

8. $15hk - 12kh + hk$

9. $80 - 80a$

10. $x^2y + 3x^2y - xy^2$

Simplifying Expressions by Using the Distributive Property

The distributive property is used when a value on the outside of parentheses is being multiplied by more than one term inside of the parentheses. For example, to simplify the expression $3(x + 2)$, the distributive property should be used to multiply 3 by x and 3 by 2 to result in the expression $3x + 6$.

However, in the expression $6(4x)$, the distributive property is not used since there is only one term inside of the parentheses. $6(4x) = 24x$.

Another example is the expression $-(x + 3)$, which is equivalent to $-x - 3$ after the distributive property is applied. The negative sign on the outside of parentheses becomes a factor of –1.

Practice
Simplify each of the following expressions.

11. $7(z + 2)$

12. $4x(3x - 1)$

13. $-2(6c)$

14. $-9(b - 4) + 4$

15. $45xy(y^2 + x)$

Translating Verbal Situations into Algebraic Expressions

Another important skill when dealing with algebraic expressions is translating various situations from words into symbolic form.

Example 1
Logan pays $30 a month for her cell phone, plus $0.05 for each minute she uses. Write an expression to represent the total cost for a month if she uses m minutes during the month.

First, the $30 a month is a once-per-month charge, so that amount will be simply added to the rest of the expression. Because Logan uses m minutes and they cost $0.05 per minute, the total amount spent on minutes is $0.05m$. Combine the two amounts to get the expression for the total cost of the month: $30 + 0.05m$.

Example 2
Joseph buys x chairs at $30 each and y tables at $100 each. Write an algebraic expression for the total amount of money he spent.

First, write an expression for the total money spent on chairs. Since there are x chairs at $30 each, this expression is $30x$. Next, write an expression for the total money spent on tables. Since there are y tables at $100 each, this expression is $100y$. To write an expression for the total Joseph spent, add the expressions to get $30x + 100y$. Keep in mind that the expression cannot be simplified further because $30x$ and $100y$ are not like terms.

Practice
Translate each of the following situations into algebraic expressions. Use the indicated letters for each expression.

16. A taxicab has an initial charge of $5 and then charges $2 for each mile, c.

17. A health club has an initiation fee of $25 and then charges $30 for each month, m.

18. An ice cream shop sells x ice cream cones at $2.50 each and y sundaes at $4.00 each.

Evaluating Algebraic Expressions for Given Values

To evaluate expressions for any given values of the variables, replace the variables with the given values and use the correct order of operations.

Example
Evaluate $xy + 10$ for $x = -2$ and $y = 4$.

First, substitute the values into the expression:

$(-2)(4) + 10$

Then, evaluate using the correct order of operations:

$-8 + 10 = 2$

Practice

Evaluate each of the following expressions for $a = 3$, $b = -5$, and $c = 0.5$.

19. $a + b$

20. $2a$

21. $4b - (a + c)$

22. $-a + 2c$

23. abc

24. $a^2 - b$

Practice

Given the formula $P = 2l + 2w$, which represents the perimeter of a rectangle for the dimensions l and w, evaluate each of the given values:

25. $l = 3$ and $w = 5$

26. $l = 0.2$ and $w = 6$

27. $l = 2.4$ and $w = 10$

Evaluating Formulas

Evaluating formulas is very similar to the previous section on evaluating expressions for specific values of the variables. In this type of problem, the formula will be given along with a value or values for some of the variables in the formula. Then, the formula can be simplified to find the missing value.

Example

In the formula $C = \frac{5}{9}(F - 32)$, F represents the temperature in degrees Fahrenheit and C represents the temperature in degrees Celsius. What is the temperature in degrees Celsius when the temperature is 50°F?

First, substitute the value of $F = 50$ into the formula:

$C = \frac{5}{9}(50 - 32)$

Follow the order of operations and subtract $50 - 32$ within the parentheses. The equation becomes $C = \frac{5}{9} \times 18$.

Use simplifying to get $C = \frac{5}{\cancel{9}_1} \times \frac{\cancel{18}^2}{1} = \frac{5 \times 2}{1}$ $= 10$. The temperature is 10°C.

Practice Answers

1. yes
2. no
3. yes
4. no
5. no
6. $21x$
7. $-8d$
8. $4hk$
9. $80 - 80a$
10. $4x^2y - xy^2$
11. $7z + 14$
12. $12x^2 - 4x$
13. $-12c$
14. $-9b + 40$
15. $45xy^3 + 45x^2y$
16. $5 + 2c$
17. $25 + 30m$
18. $2.50x + 4.00y$
19. -2
20. 6
21. -23.5
22. -2
23. -7.5
24. 14
25. 16
26. 12.4
27. 24.8

LINEAR EQUATIONS AND WORD PROBLEMS
For the COMPASS, ASSET, and ACCUPLACER

Any intelligent fool can make things bigger, more complex, and more violent. It takes a touch of genius—and a lot of courage— to move in the opposite direction.

—ALBERT EINSTEIN

This chapter covers a lot of ground, and might take you longer than 20 minutes to complete. Split up your study into two parts if you need extra time to make your way through this lesson's examples.

LESSON SUMMARY
Linear equation solving is an important skill to master, as it forms the basis for advanced algebra. While it's true that simple equations can often be solved with mental math, just as you usually learn to swim in shallow water before you venture into the deep, you should learn the skills of equation solving with simple equations. Then, when the opportunity arises to solve a more complex problem, you will be very fluent in your skills. This lesson reviews solving all sorts of linear equations, including two-step equations, equations that require simplification by the distributive property or combining like terms, and equations with variables on both sides of the equation. In addition, this lesson demonstrates how to take a word problem and conquer it by writing and solving an equation.

Equations

Lesson 7 reviewed the concept of algebraic expressions. An equation is a mathematical sentence that states that two expressions are equal. At least one of the expressions is a variable expression—it contains a variable. The goal of solving an equation is generally to get an answer. This is done by isolating the variable, which is getting the variable alone on one side of the equation. While you are in the process of solving, keep this goal in mind. It will help you to make correct decisions.

To isolate the variable, you first simplify each side of the equation where possible, and then perform opposite operations. When you see addition, you subtract this term from both sides. When you see multiplication, you divide the coefficient from both sides.

QUICK FACTS

The opposite of addition is subtraction.
The opposite of multiplication is division.
The opposite of squaring is to take the square root.

Solving One-Step and Two-Step Equations

The simplest equations to solve involve one or two steps to isolate the variable.

Example 1

For example, to solve $x + 10 = 24$, you do the opposite operation of "plus 10" to both sides of the equation:

$$x + 10 = 24$$
$$\underline{-10 = -10}$$ Do the opposite of +10 to both sides.

$$x = 14$$ The variable is isolated, so the equation is solved.

A two-step equation takes two steps to solve. The two steps are usually addition or subtraction, and then either multiplication or division. When you isolate the variable, remember that you are working backward. When you perform opposite operations, you work backward in the order of operations—that means you do the opposite of addition or subtraction first, and then the opposite of multiplication and division.

Example 2

Solve $5m - 12 = 18$

$$5m - 12 = 18$$
$$\underline{+12 = +12}$$ Add 12 to both sides (the
$$5m = 30$$ opposite of subtracting 12) (first step).
$$\frac{5m}{5} = \frac{30}{5}$$ Divide by 5 on both sides (the opposite of multiplication) (second step).
$$m = 6$$ The variable is isolated, so the equation is solved.

If one of the operations is division, as shown by a fraction bar, you undo that operation by multiplication.

Example 3

$$\frac{x}{6} + 10 = 14$$
$$-10 = -10$$ Subtract 10 from both sides (the opposite of adding 10) (first step).

$$\frac{x}{6} = 4$$
$$6 \times \frac{x}{6} = 4 \times 6$$ Multiply by 6 on both sides (the opposite of dividing by 6) (second step).
$$x = 24$$ The variable is isolated, so the equation is solved.

TIP

Any equation that you solve can be checked to be sure you solved it correctly. Take the answer you arrived at, and substitute into the original equation. You reviewed substitution in Lesson 7. Simplify each side of the equation, and if the two sides are equal, then you most likely have solved the equation correctly! For example, in the previous equation, you solved for x and determined its value to be 24. Substitute 24 into $\frac{x}{6} + 10 = 14$ to get $\frac{24}{6} + 10 = 14$ or $4 + 10 = 14$. You can verify that $14 = 14$, so you know you solved the equation correctly.

You can even use this technique on a multiple-choice section of a test. If you are stuck and do not know how to proceed to solve, you can substitute each answer choice into the equation to find the choice that makes the equation true.

Practice

Solve:

1. $-5x = 45$

2. $a - 18 = 12$

3. $4m + 6 = 42$

4. $-7x - 3 = -52$

5. $\frac{x}{4} + 2 = 5$

The Distributive Property and Equations

Advanced equations are those with more than two steps. There are several steps that are performed before you undo addition or subtraction, and also multiplication and division. The first step, if appropriate, is to simplify using the distributive property. You reviewed the distributive property in Lesson 7, Algebraic Expressions.

Example

For example, to solve $4(p + 3) = 30$, the first step is to distribute the 4 to both terms in parentheses:

$$4(p + 3) = 30$$
$$4p + 12 = 30 \quad \text{Once you have distributed, you have a two-step equation.}$$
$$\underline{-12 = -12} \quad \text{Subtract 12 from both}$$
$$4p = 18 \quad \text{sides (the opposite of adding 12).}$$
$$\frac{4p}{4} = \frac{18}{4} \quad \text{Divide both sides by 4 (the opposite of multiplying by 4).}$$
$$p = 4.5 \quad \text{The variable is isolated, so the equation is solved.}$$

Do not forget that each and every equation can be checked. To check this, substitute in 4.5 for p in the original equation: $4(4.5 + 3) = 30$, or $4(7.5) = 30$ and $30 = 30$; the solution is correct.

Sometimes you are distributing a negative term:

$$-2(x - 4) = -20$$

You distribute the -2 to the x term and to the -4 term to get:

$$-2x + 8 = -20 \quad \text{This is now a two-step equation.}$$
$$\underline{-8 \quad -8} \quad \text{Subtract 8 from both sides (the}$$
$$-2x = -28 \quad \text{opposite of adding 8).}$$
$$\frac{-2x}{-2} = \frac{-28}{-2} \quad \text{Divide both sides by } -2 \text{ (the opposite of multiplying by } -2\text{).}$$
$$x = 14 \quad \text{The variable is isolated, so the equation is solved.}$$

Practice

Solve:

6. $5(y + 3) = 37.5$

7. $-4(x + 8) = 20$

8. $-6(n - 7) = -60$

Solving Equations That Require Combining Like Terms

Another step used in simplifying when solving an equation is to combine like terms. You reviewed this concept in Lesson 7.

Example 1

To solve $7x - 23 - 10x = 7$, before you start to do opposite operations to both sides you must simplify by combining the like terms of $7x$ and $-10x$:

$$7x - 23 - 10x = 7$$

$-3x - 23 = 7$ Combine $7x$ and $-10x$.

$\underline{+ 23 = +23}$ Add 23 to both sides (the

$-3x = 30$ opposite of subtracting 23).

$\frac{-3x}{-3} = \frac{30}{-3}$ Divide both sides by -3 (the opposite of multiplying by -3).

$x = -10$ The variable is isolated, so the equation is solved.

Remember that every equation can be checked. For this one, substitute in -10 for x to get $7(-10) - 23 - 10(-10) = 7$, or $-70 - 23 + 100 = 7$. This simplifies to $-93 + 100 = 7$. The solution is correct.

Example 2

To solve $\frac{1}{4}x - 10 + \frac{x}{4} = 14$, recognize that $\frac{x}{4}$ is another way to write $\frac{1}{4}x$, and add $\frac{1}{4}x$ and $\frac{1}{4}x$ to get $\frac{1}{2}x$.

$\frac{1}{2}x - 10 = 14$ Combine like terms.

$\underline{+ 10 = + 10}$ Add 10 to both sides (the

$\frac{1}{2}x = 24$ opposite of subtracting 10).

$2 \times \frac{1}{2}x = 24 \times 2$ Multiply both sides by 2 (the opposite of multiplying by $\frac{1}{2}$).

$x = 48$ The variable is isolated, so the equation is solved.

You may have an equation where it is necessary to both distribute and combine like terms. In this case, do the distributive property first, since multiplication comes before addition and subtraction in the order of operations.

Example 3

To solve $4(s - 2) + 6s = -38$:

$4s - 8 + 6s = -38$ Do the distributive property.

$10s - 8 = -38$ Combine like terms.

$\underline{+ 8 = + 8}$ Add 8 to both sides (the

$10s = -30$ opposite of subtracting 8).

$\frac{10s}{10} = \frac{-30}{10}$ Divide both sides by 10 (the opposite of multiplying by 10).

$s = -3$ The variable is isolated, so the equation is solved.

Practice

Solve:

9. $-12t + 20 + 4t = -12$

10. $\frac{x}{6} - 3 + \frac{x}{6} = 15$

11. $10(m + 4) + 2m = 160$

Solving Equations with a Variable on Both Sides

You have reviewed the two simplifying steps that are done first when solving advanced equations. These operations are done to each side of the equation separately. A solving step, which involves inverse operations, is when there is a variable on both sides of the equation. This is done before any two-step operations.

Example 1

To solve $25y - 15 = 13y + 9$, first do an inverse operation (addition or subtraction) to get the variable on one side of the equation. It is advisable to undo the smaller variable term because it avoids negative coefficients.

$$
\begin{array}{ll}
25y - 15 = 13y + 9 & \\
\underline{- 13y = -13y} & \text{Subtract } 13y \text{ from each side (to eliminate } 13y \text{ on the right side).} \\
12y - 15 = 9 & \text{Now there is a two-step equation.} \\
\underline{+15 = +15} & \text{Add 15 to both sides (step one of the two-step equation part).} \\
12y = 24 & \\
\frac{12y}{12} = \frac{24}{12} & \text{Divide both sides by 12 (step two of the two-step equation part).} \\
y = 2 & \text{The variable is isolated, so the equation is solved.}
\end{array}
$$

Remember that you always do the distributive property first, if applicable.

Example 2

To solve $-3(x - 7) = x - 7$:

$$
\begin{array}{ll}
-3x + 21 = x - 7 & \text{Distribute the } -3 \text{ first.} \\
\underline{+3x = +3x} & \text{Add } 3x \text{ to both sides (eliminate the smaller variable term).} \\
21 = 4x - 7 & \\
\underline{+7 = +7} & \text{Add 7 to both sides of the equation (the opposite of subtracting 7).} \\
28 = 4x & \\
\frac{28}{4} = \frac{4x}{4} & \text{Divide both sides by 4 (the opposite of multiplying by 4).} \\
7 = x & \text{The variable is isolated, so the equation is solved.}
\end{array}
$$

Just as in all equations, you can check your answer. In this case, substitute 7 for x in the original equation to get $-3(7 - 7) = 7 - 7$, or $-3(0) = 0$, or $0 = 0$. The two sides are equal, so the solution is correct.

Practice

Solve:

12. $9x - 16 = 4x + 24$

13. $-q + 3 = 6q + 10$

14. $3x - 15 + 8x = 5x + 15$

15. $-4(m - 3) = 3m - 9$

TIP

When solving advanced equations, follow this order:

1. Do the distributive property, if applicable, to each side of the equation separately.
2. Combine like terms, if applicable, on each side of the equation separately.
3. If there is a variable on both sides, eliminate the smaller variable term through addition or subtraction.
4. Undo addition or subtraction of the constant term.
5. Undo multiplication or division of the coefficient.
6. Finally, circle your answer and check to see if your solution works.

Applications

Equations can be used to problem solve. In Lesson 7, you translated expressions. You can use that same skill to translate words into equations.

Example 1

For example, "twelve less than five times a number is eighteen" is translated to $5n - 12 = 18$, and this can be solved:

$$5n - 12 = 18$$
$$\underline{+12 = +12} \quad \text{Add 12 to both sides of the}$$
$$5n = 30 \quad \text{equation (the opposite of subtracting 12).}$$
$$\frac{5n}{5} = \frac{30}{5} \quad \text{Divide both sides by 5 (the opposite of multiplying by 5).}$$
$$n = 6 \quad \text{The variable is isolated, so the equation is solved.}$$

Sometimes, the measures of the sides of a polygon are algebraic expressions, and you need to solve an equation to find the measures of the sides.

Example 2

For example, if the perimeter of the following rectangle is 66 cm, what are the value of x and the value of the sides?

$x + 8$ cm

$2x + 4$ cm

Using the facts that the sum of the measures of the sides of a polygon is the perimeter, and that opposite sides of a rectangle are congruent, set up an equation:

$$2(2x + 4) + 2(x + 8) = 66 \quad \text{Set up the equation.}$$
$$4x + 8 + 2x + 16 = 66 \quad \text{Distribute both 2's.}$$
$$6x + 24 = 66 \quad \text{Combine like terms.}$$
$$\underline{-24 = -24} \quad \text{Subtract 24 from}$$
$$6x = 42 \quad \text{both sides (the opposite of adding 24).}$$
$$\frac{6x}{6} = \frac{42}{6} \quad \text{Divide both sides by 6 (the opposite of multiplying by 6).}$$

$$x = 7 \quad \text{The variable is isolated, so the equation is solved.}$$

Once you have solved for the value of x, you can substitute this value in to determine the measure of the sides. The shorter side is $x + 8$, or $7 + 8 = 15$. The longer side is $2x + 4 = 2(7) + 4 = 14 + 4 = 18$. Check to verify that $15 + 15 + 18 + 18 = 66$.

You can solve real-life problems with algebraic equations.

Example 3

Consider this problem:

Anika buys fudge at the wholesale price of $2.95 per pound. She sells the fudge for $8.95 per pound. Her monthly expenses for rent and utilities are $320. How many pounds of fudge must she sell each month to make a monthly profit of $1,000?

This can be solved by letting a variable such as n represent the number of pounds of fudge Anika must sell each month. Profit is income minus expenses. So her profit P is represented by $P = n(8.95 - 2.95) - 320$. To find the number of pounds, solve:

$$1,000 = n(8.95 - 2.95) - 320$$
$$1,000 = 6n - 320 \quad \text{Do the subtraction within the parentheses.}$$
$$\underline{+320 = +320} \quad \text{Add 320 to each}$$
$$1,320 = 6n \quad \text{side (the opposite of subtracting 320).}$$
$$\frac{1,320}{6} = \frac{6n}{6} \quad \text{Divide both sides by 6 (the opposite of multiplying by 6).}$$
$$220 = n \quad \text{She has to sell 220 pounds of fudge.}$$

Practice

For the following problems, write an equation and solve.

16. Sixteen less than three times a number is equal to the number plus 10. What is the number?

17. The following triangle is an isosceles triangle with a perimeter of 48 cm. What is the value of x and the length of each of the sides?

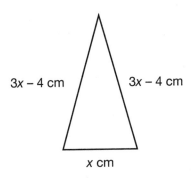

18. A fitness facility charges $33 per month and $6 for each spinning class. How many spinning classes can you attend each month to keep your monthly charge at $105?

19. Bicycles are bought wholesale for $23 each. They are sold at the bicycle store for $235 each. The store has monthly expenses of $664 for rent and utilities. How many bikes must be sold to have a profit of $4,000 per month?

Solving an Equation in Terms of a Variable

Later, in Lesson 13, you will learn how to solve simultaneous equations with more than one variable. If you have only one equation, but two variables, you can solve for one variable in terms of another. When you do this, you will not get a numerical answer, but a variable expression. Just remember that the goal is to isolate the variable that you are asked to solve for.

Example 1

You can solve the equation $y = 20x + 12$ for x, in terms of y:

$$\begin{array}{rl} y &= 20x + 12 \\ -12 &= \quad\quad -12 \\ \hline y - 12 &= 20x \\ \frac{y-12}{20} &= \frac{20x}{20} \\ \frac{y-12}{20} &= x \end{array}$$

To isolate x, first subtract 12 from each side.

The second step is to divide both sides by 20.

The variable x is isolated, so the equation is solved.

You can also use substitution with various equations to solve for one variable in terms of another.

Example 2

If given that $w = 5x$, and also that $35x = 2z$, you can solve for z in terms of w.

$w = 5x$ — Given.
$7w = 35x$ — Multiply both sides by 7.
$35x = 2z$ — Given.
$7w = 2z$ — Substitute $2z$ for $35x$.
$\frac{7w}{2} = \frac{2z}{2}$ — Divide both sides by 2 (the opposite of multiplying by 2).
$\frac{7w}{2} = z$ — The variable z is isolated, so the equation is solved.

Practice

Solve for one variable in terms of another.

20. Solve for n in terms of p: $p = -4n + 8$

21. Solve for x in terms of y: $10y = 5x - 20$

22. Given that $a = 4b$ and $24b = 32c$, solve for c in terms of a.

23. Given that $2w = 12x$ and $6x = 6y$, solve for y in terms of w.

Practice Answers

1. $x = -9$
2. $a = 30$
3. $m = 9$
4. $x = 7$
5. $x = 12$
6. $y = 4.5$
7. $x = -13$
8. $n = 17$
9. $t = 4$
10. $x = 54$
11. $m = 10$
12. $x = 8$
13. $q = -1$
14. $x = 5$
15. $m = 3$
16. The equation is $3n - 16 = n + 10$. The number is $n = 13$.
17. The equation is $x + (3x - 4) + (3x - 4) = 48$; $x = 8$. The side lengths are 20 cm, 20 cm, and 8 cm.
18. The equation is $6s + 33 = 105$. The number of spinning classes per month, s, is 12.
19. The equation is $b(235 - 23) - 664 = 4{,}000$. The number of bicycles, b, is 22.
20. $\frac{p-8}{-4} = n$
21. $x = 2y + 4$
22. $\frac{6a}{32} = c$ or $\frac{3a}{16} = c$
23. $\frac{2w}{12} = y$ or $\frac{w}{6} = y$

9 ▶ INEQUALITIES
For the ASSET and ACCUPLACER

One person's constant is another person's variable.
—SUSAN GERHART

LESSON SUMMARY
Using constants and variables is essential to the study of equations and inequalities. The process of working with inequalities is very similar to equations, with a few differences. This lesson covers those differences, which include the types of symbols used and cases where the direction of the symbol changes. You will also learn how to translate verbal statements into symbolic form, graph inequality solution sets, and solve one-step and multistep inequalities.

 s the study of inequalities begins, first let's identify the basic inequality symbols.

Translating Inequalities

An important skill in algebra is to be able to take verbal statements and change them into symbols, numerals, and operations. In order to translate from verbal phrases into these mathematical symbols, let's review some of the basic phrases of inequality. See the following chart for a summary.

$<$	$>$	$\leq$	$\geq$
less than smaller than	greater than more than	less than or equal to at most not greater than the greatest value the maximum value	greater than or equal to at least not less than the least value the minimum value

Here are a few examples of verbal phrases translated into mathematical symbols:

> **Example 1:** The phrase *x is greater than or equal to 10* translates to $x \geq 10$.
> **Example 2:** The phrase *g is less than five* translates to $g < 5$.
> **Example 3:** The phrase *four more than n is at least 12* translates to $n + 4 \geq 12$.
> **Example 4:** The phrase *the maximum value of x is 25* translates to $x \leq 25$.

Practice

Translate each of the following statements into symbolic form. Use $>$, $\geq$, $<$, $\leq$, or $\neq$.

1. A number x is less than -5.

2. Twice the value of n is greater than 30.

3. The value of n is not equal to 8.

4. The variable y added to 15 is at most 46.

5. The value of c is at least 17.

6. The value of k is not more than 9.5.

7. Seven more than b is not greater than -10.

8. One-half of x has a minimum value of 14.

Graphing Inequalities

When solving inequalities, a graph cam be used to show an entire solution set since the solution usually contains multiple correct answers. For example, if x is less than 5 ($x < 5$), then all numerical values smaller than 5 are solutions. To show this set, a ray is drawn on a number line to illustrate all possible answers. For this particular example, a ray is drawn beginning at 5 and continuing through all values to the left of 5 on the number line. An open circle is used over the 5 since 5 is not greater than itself and not included in the solution.

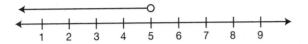

For the example $3.5 \geq x$, the graph would have a closed (filled-in) circle at 3.5 (since 3.5 *is* less than or equal to 3.5) with the arrow pointing to the left.

For the inequality $y > 10$, the graph would have an open circle at 10 with the arrow pointing to the right.

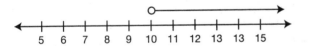

The graph of $n \leq -1$ has a closed circle at -1 with the arrow pointing to the left.

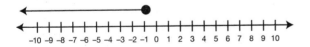

TIP

When graphing > or <, use an open circle.
When graphing ≥ or ≤, use a closed (filled-in) circle.

Practice

Graph each of the following inequalities on a number line.

9. $x \geq 6$

10. $y < -3$

11. $t > -9$

12. $m \leq 12$

13. $n < 0$

14. $a > -1$

Solving One-Step Inequalities

To solve one-step inequalities, follow the steps for solving one-step equations from Lesson 8. For each problem, perform the inverse operation to get the variable alone.

Example 1
Solve $x + 7 > 14$.

In this problem, subtract 7 from each side of the inequality to perform the inverse operation.

The inequality becomes $x + 7 > 14$

$$\underline{-7 \quad -7}$$

This simplifies to $\quad x > \quad 7$

Example 2
Solve $y - 16 \leq -10$.

In this inequality, add 16 to each side to get the variable alone.

The inequality becomes:

$y - 16 \leq -10$

$\underline{+16 \quad +16}$

$y \quad\quad \leq \quad 6$

QUICK FACT

When multiplying or dividing each side of an inequality by a negative value, switch the direction of the inequality symbol. When the signs of both sides of an inequality change, the values become their opposites. Thus, what was smaller before is now larger and what was larger is now smaller.

Solving Multistep Inequalities

As with equations, perform the inverse operations to get the variable alone, reversing the order of operations. Review the steps for solving multistep equations from Lesson 8 as needed.

Example 1

Solve $5x - 3 > 12$.

First, add 3 to each side of the inequality to get

$$5x - 3 > 12$$
$$\underline{ + 3 \quad + 3}$$

The inequality simplifies to $5x > 15$
Divide each side by 5 to get $\frac{5x}{5} > \frac{15}{5}$
This simplifies to $x > 3$.

Example 2

Solve $-2(3x - 5) \geq 40$.

First, use the distributive property to simplify the left side of the inequality.

The inequality becomes $-6x + 10 \geq 40$.
Subtract 10 from each side of the inequality.

The inequality becomes $\underline{-6x + 10 \geq 40}$
$$ - 10 \quad\;\; - 10$$

which simplifies to $-6x \geq 30$
Divide each side of the inequality by -6, and switch the direction of the inequality symbol.

$$\frac{-6x \geq 30}{-6 \quad\;\; -6}$$

The solution to the inequality is $x \leq -5$.

Example 3

Solve $-5x > 25$.

To get the variable alone in this inequality, divide each side of the inequality by -5. Because you are dividing by a negative value, switch the direction of the inequality symbol. The inequality becomes $\frac{-5x}{-5} > \frac{25}{-5}$, which simplifies to $x < -5$.

Because the first step in solving this inequality was to divide each side by -5, the direction of the inequality symbol needed to be switched. The inequality contained a *greater than* symbol, but following the step $\frac{-5x}{-5} > \frac{25}{-5}$, the symbol becomes *less than*: $x < -5$.

Example 4

Solve $\frac{y}{4} \geq 8$.

Multiply each side of the inequality by 4 to get the variable alone. The inequality becomes $\frac{y}{4} \times 4 \geq 8 \times 4$, and simplifies to $y \geq 32$.

Practice

Solve each of the following inequalities.

15. $3x > -9$

16. $y - 12 \leq -5$

17. $n + 8 < 0$

18. $\frac{r}{-2} \geq -10$

Practice

Solve each of the following inequalities.

19. $4x + 1 > -7$

20. $-6m - 15 < 15$

21. $3(y - 12) \leq -3$

22. $-\frac{1}{2}(n + 14) - 2 \leq 8$

23. $\frac{r}{-7} - 20 \geq -15$

24. $5x + 10 \geq 28 + 3x$

DID YOU KNOW?

When a variable is located between two values, two inequality symbols are used. This is known as a **compound inequality**. For example, in order to say that a value x is between 10 and 20, two less than symbols are used and the variable x is placed between the values. The inequality is $10 < x < 20$. This translates to "10 is less than x, which is less than 20."

Absolute Value Inequalities

Absolute value, as explained in Lesson 2, may also be incorporated into inequality problems. Take, for example, the absolute value inequality $|x| < 2$. Recall that in the equation $|x| = 2$, the solution becomes two equations; $x = 2$ or $x = -2$. So for this inequality, all values between -2 and 2 have an absolute value that is less than 2. Thus, the solution set is the compound inequality $-2 < x < 2$.

If the variable in the inequality is *greater than* a value, then the solutions are greater than the given value, or smaller than the opposite of the value. For example, for the inequality $|x| > 6$, the values in the solution set are greater than 6 or less than -6. This is represented by the inequality $x < -6$ or $x > 6$.

For an additional example, take the inequality $|x + 3| \leq 5$. To solve this problem, consider all possible values of x where the expression $x + 3$ either includes or is between -5 and 5. This becomes the compound inequality $-5 \leq x + 3 \leq 5$. Subtract 3 from all three parts of the inequality to get $-5 - 3 \leq x + 3 - 3 \leq 5 - 3$. This simplifies to the compound inequality $-8 \leq x \leq 2$.

Practice
Solve each of the following inequalities. Express the solution set as a compound inequality.

25. $|x| < 10$

26. $|y| > 4$

27. $|m - 8| \leq 12$

Applications

In order to solve word problems that incorporate inequalities, use the key words and phrases of inequalities studied at the start of this lesson. Then, set up the inequality and solve using the steps practiced in the previous sections.

Example 1
The combined weight of Kevin's two dogs is at most 78 pounds. If the weight of one of the dogs is 26 pounds, what is the maximum weight of the other dog?

The key phrase *at most* represents the relationship less than or equal to. Set up the inequality $w + 26 \leq 78$, where w represents the weight of the other dog. Subtract 26 from each side of the inequality to get $w + 26 - 26 \leq 78 - 26$, which simplifies to $w \leq 52$. The maximum weight of the other dog is 52 pounds.

Example 2
Charlene earns $12 per hour at her job. If she needs to earn at least $350 for the week, what is the minimum number of hours she should work this week?

The key phrase *at least* represents the relationship greater than or equal to. Set up the inequality $12h \geq 350$, where h represents the number of hours that Charlene should work to make $350 or more. Divide each side of the inequality by 12 to get the inequality $\frac{12h}{12} \geq \frac{350}{12}$, which simplifies to $h \geq 29.1\overline{6}$. This solution needs to be rounded up since 29 hours would not give a total of $350 or more. Therefore, Charlene needs to work at least 30 hours.

Practice

Write and solve an inequality for each of the following.

28. JoAnn buys a movie ticket for $9.25. If she has a total of $18, how many snacks can she purchase if they are $2.75 each?

29. Henry's car gets at the most 25 miles per gallon, and his gas tank holds 12 gallons. How many times will he need to fill up his car on a trip of 1,250 miles if he begins with a full tank?

30. Teddy has eaten a total of 1,675 calories for the day so far. If he would like to eat a maximum of 2,200 calories for the day, what is the greatest number of calories he should consume for the remainder of the day?

31. A restaurant serves 400 servings of coffee on an average day. If a can of coffee makes approximately 75 servings, what is the least number of cans of coffee that the restaurant needs for an average day?

32. Shelly is selling T-shirts for a fund-raiser, and she plans on selling 125 shirts. If her cost from the factory is $6.50 per shirt, what is the minimum price she should charge for each shirt if she would like to raise at least $200?

33. In order to seed her garden, Julia needs a minimum of one pound of seeds for every 300 square feet. If her garden has a total of 1,675 square feet, what is the least number of pounds of seeds that she needs?

Practice Answers

1. $x < -5$

2. $2n > 30$

3. $n \neq 8$

4. $15 + y \leq 46$

5. $c \geq 17$

6. $k \leq 9.5$

7. $b + 7 \leq -10$

8. $\frac{1}{2}x \geq 14$

9.

10.

11.

12.

13.

14.

15. $x > -3$

16. $y \leq 7$

17. $n < -8$

18. $r \leq 20$

19. $x > -2$

20. $m > -5$

21. $y \leq 11$

22. $n \geq -34$

23. $r \leq -35$

24. $x \geq 9$

25. $-10 < x < 10$

26. $y < -4$ or $y > 4$

27. $-4 \leq m \leq 20$

28. $9.25 + 2.75x \leq \$18$; 3 snacks

29. $25(12)(x - 1) \geq 1{,}250$; 4 times

30. $1{,}675 + x \leq 2{,}200$; 525 calories

31. $75x \geq 400$; 6 cans

32. $125(x - 6.50) \geq 200$; \$8.10 per T-shirt

33. $300x \geq 1{,}675$; 6 pounds

10 ▶ BASIC GEOMETRY
For the COMPASS, ASSET, and ACCUPLACER

And since geometry is the right foundation of all painting, I have decided to teach its rudiments and principles to all youngsters eager for art.

—ALBRECHT DURER

LESSON SUMMARY

The word *geometry* means measure of the earth. This branch of mathematics is concerned with the shape, size, and characteristics of two- and three-dimensional figures. This lesson reviews the basic concepts of angles, triangles, quadrilaterals, and circles. You will learn about the concepts of measurement as they relate to two- and three-dimensional figures.

Angles

Angles are formed when two rays share the same endpoint, called the vertex. Angles are named with the vertex in the middle. For example, the angle shown in the following figure is angle $\angle ABC$, or it can be named $\angle CBA$. It is formed by the rays $\overrightarrow{AB}$ and $\overrightarrow{BC}$.

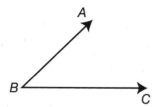

Because point *B* is the vertex, it must be named in the middle. Angles are measured by a protractor, in units called degrees, and are classified according to their measure. *Congruent* angles have the same measure. The notation for the measure of an angle is *m∠*. For example, *m∠ABC* = 28° states that the measure of angle *ABC* is 28°.

QUICK FACTS

An *acute angle* is an angle with measure greater than 0° but less than 90°.

A *right angle* is an angle with a measure of exactly 90°.

An *obtuse angle* is an angle with measure greater than 90° but less than 180°.

A *straight angle* is an angle with a measure of exactly 180°.

A *reflex angle* is an angle with measure greater than 180° but less than 360°.

Practice
Identify the type of angle based on the measure.

1. An angle of 47°

2. An angle of 190°

3. An angle of 90°

4. An angle of 100°

5. An angle of 180°

Special Angle Pairs

There are special angle pairs that are frequently used and referred to in geometry. The first two are complementary and supplementary angles.

DEFINITIONS

Complementary angles are a pair of angles whose angle measures add to 90°.
Supplementary angles are a pair of angles whose angle measures add to 180°.

When two angles have measures that add to 90° they are called complementary, or complements of each other. Supplementary angles are two angles that have measures that add to 180°; they are called supplements of each other. These angles may or may not be adjacent to each other. When complementary angles are adjacent, they form a right angle; when supplementary angles are adjacent, they form a straight angle and are called a linear pair.

Practice
Find the complement and supplement of the given angles.

6. 62°

7. 18°

8. 84°

9. Find the missing angles *x* and *y* in the figure.

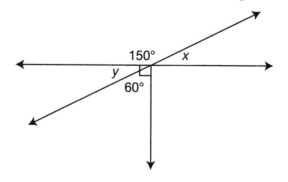

Vertical Angles

Vertical angles are another special pair of angles. Vertical angles are angles formed by two intersecting lines. They are opposite each other; they share a vertex and no other points in common. In the following figure, ∠1 and ∠3 are vertical angles, and ∠2 and ∠4 are vertical angles. Vertical angles are congruent; that is, they have the same measure.

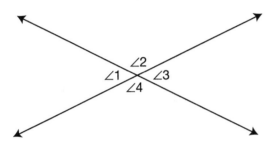

Note also that in the figure there are four pairs of supplementary angles: ∠1 and ∠2, ∠1 and ∠4, ∠2 and ∠3, and ∠3 and ∠4.

Practice

10. Find the missing angle *x* in this figure:

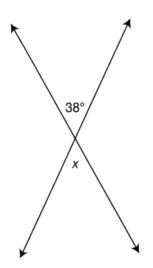

For the following questions, find the value of *m* and then the measure of the angle, or angles, if applicable.

11.

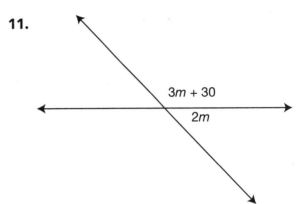

12.

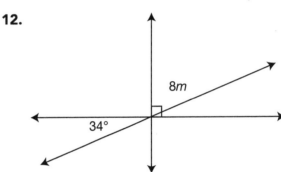

Angle Pairs Formed When a Transversal Line Crosses Two Parallel Lines

When a line called a *transversal line* crosses two parallel lines, eight angles are formed, as shown:

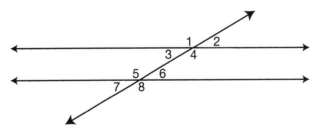

There are many special angle pairs. Two pairs you already reviewed are vertical angles (such as ∠2 and ∠3) and supplementary angles (such as ∠2 and ∠4). There are three other special angle pairs that all form pairs of angles that are congruent—they have the same angle measure. *Alternate interior angles* are angles in the interior of the parallel lines (that is, between them), on opposite sides of the transversal. These angles shown in the figure are ∠3 and ∠6, and another pair, ∠4 and ∠5. *Alternate exterior angles* are outside the two parallel lines and on opposite sides of the transversal. These angles are shown in the figure as one pair, ∠1 and ∠8, and the other pair, ∠2 and ∠7. *Corresponding angles* are angle pairs are on the same side of the transversal—one in the interior and one in the exterior. These are shown in the figure as ∠1 and ∠5, ∠3 and ∠7, ∠2 and ∠6, and ∠4 and ∠8. Just as with the other angle pairs described in the preceding section, you can solve for either missing variables or missing angle measures.

Practice

Find the missing angles x, y, and z in this figure:

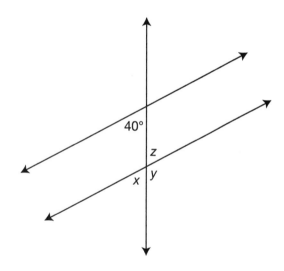

13. Angle x

14. Angle y

15. Angle z

For the following questions, find the value of p and then the measure of the angles.

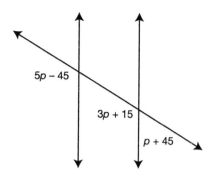

16. $5p - 45$

17. $3p + 15$

18. $p + 45$

Triangles

Triangles are three-sided closed geometric figures. They consist of three sides and three angles. They are classified two ways: according to angles and sides.

QUICK FACTS

Triangles classified by angles:

An *acute triangle* is a triangle with three acute angles.

A *right triangle* is a triangle with a right angle.

An *obtuse triangle* is a triangle with an obtuse angle.

Triangles classified by sides:

A *scalene triangle* is a triangle with sides of different lengths.

An *isosceles triangle* is a triangle with two sides of the same length.

An *equilateral triangle* is a triangle with all three sides the same length. All the angles in an equilateral triangle measure 60°.

Look, for example, at these triangles:

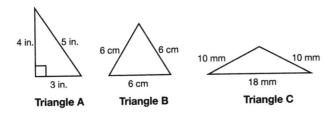

Triangle A   **Triangle B**   **Triangle C**

Triangle A is a right scalene triangle. It has one right triangle, and all the sides are different lengths. Triangle B is an acute equilateral triangle. All of the sides are congruent and the angles are all acute. Triangle C is an obtuse isosceles triangle. One of the angles is obtuse, and two of the sides are congruent.

Practice

Classify the following triangles by both angles and sides.

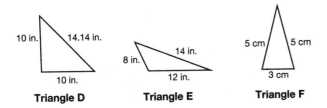

Triangle D   **Triangle E**   **Triangle F**

19. Triangle D

20. Triangle E

21. Triangle F

Angle Measure of a Triangle

The sum of the measures of the angles of a triangle is 180°. An *exterior angle* to a triangle is an angle that is a linear pair to any one of the angles in the triangle. The measure of an exterior angle in a triangle is equal to the sum of the measures of the other two angles in

the triangle (the angles that are not linear pairs to this exterior angle).

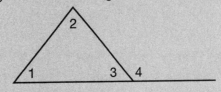

QUICK FACTS

The following figure shows the angles in a triangle with exterior angle $\angle 4$.

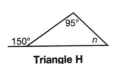

$m\angle 1 + m\angle 2 + m\angle 3 = 180°$
$m\angle 3 + m\angle 4 = 180°$
$m\angle 4 = m\angle 1 + m\angle 2$

Practice

Use the following figures to find the missing angles.

Triangle G   **Triangle H**   **Triangle J**

22. Angle x in Triangle G

23. Angle n in Triangle H

24. Angle p in Triangle J

Use the following figure to find the given angle measures. Lines m and n are parallel.

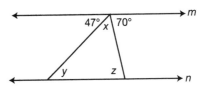

25. Angle x

26. Angle y

27. Angle z

Quadrilaterals

Quadrilaterals are four-sided closed figures. There is also a classification system for quadrilaterals. The graphic organizer that follows shows the classification and the facts associated with each type of quadrilateral. For example, a parallelogram is a special quadrilateral, and a trapezoid is a special quadrilateral. Note that a square is both a rectangle and a rhombus, and therefore shares all the characteristics of the figures shown above it in the diagram. Thus, a square is a rectangle, a rhombus, a parallelogram, and a quadrilateral.

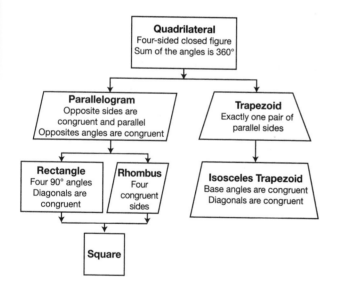

Answer true or false.

28. All rectangles are parallelograms. _____

29. Rectangles have diagonals that are congruent. _____

30. All parallelograms are rectangles. _____

31. All rectangles are squares. _____

32. Trapezoids have one pair of parallel sides. _____

33. Parallelograms have opposite sides that are both congruent and parallel. _____

Angles in Quadrilaterals

The sum of the measures of the angles in a quadrilateral is 360°. You may also encounter exterior angles with a quadrilateral. They are still a linear pair with one of the angles in the quadrilateral. Because parallelograms have parallel sides, you may encounter problems that deal with a parallelogram's parallel lines and special angle pairs. For example, look at the following figure. If it is given that the figure is a parallelogram, then angle x is 127°. The interior angle that is a linear pair with the 53° marked is $180° - 53° = 127°$. Opposite angles in a parallelogram are congruent, so $m\angle x = 127°$.

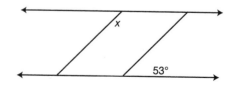

Practice

Find the value of the variable(s) and the measure of the angles.

34.

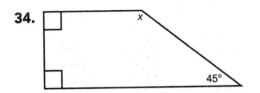

35.

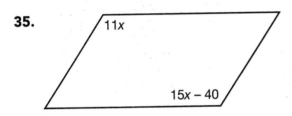

36.

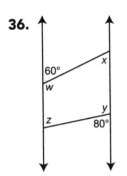

Lines of Symmetry

Some geometric figures are symmetric. The ones that are symmetric have a *line of symmetry*, which divides the figure into two congruent halves that are mirror images. Imagine that the line is a fold line and if you would fold on this line, the figure would fit on top of itself. Some figures may have several lines of symmetry. The following figures are a triangle, a quadrilateral, and a pentagon that have lines of symmetry.

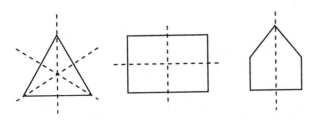

Practice

Indicate whether the following figures have one or more lines of symmetry. If they do have symmetry, tell how many lines of symmetry there are.

37. Isosceles triangle

38. Square

39. Parallelogram that is not a rhombus or a rectangle

Circles

A circle is defined by all of the points that are the same distance from a fixed point called the center.

DEFINITIONS

A *radius* is a line segment whose endpoints are the center of the circle and any point on the circumference of the circle.

A *diameter* is a line segment whose endpoints are points on the circle and that passes through the center.

The length of the diameter is twice the length of the radius.

A *unit circle* is a circle with a radius of 1 unit.

A *central angle* is an angle in a circle whose vertex is the center of the circle.

An *arc* is a portion of the circumference (the perimeter) of a circle.

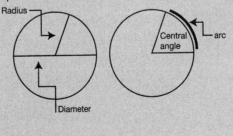

Angles in a Circle

Angles in a circle are measured by degrees or radians. Pi (π) radians are equivalent to 180°. A full circle has 360°, or 2π radians. Pieces of the circle, bounded by the sides of a central angle, can be called wedges, and they are parts of the whole. Use percentages and proportions to determine the angle measure of the wedges.

Example

For example, 25% of the circle is a wedge with a central angle of 90°, because $\frac{25}{100} = \frac{90}{360}$. To find the angle measure for 30% of the circle, solve the proportion $\frac{30}{100} = \frac{x}{360}$, and cross multiply to find that the central angle is 108°. To convert 108° to radians, use the proportion $\frac{108}{x} = \frac{180}{\pi}$. Cross multiply to get $108\pi = 180x$. Divide by 180 and $108° = \frac{3\pi}{5}$.

Arcs in a Circle

Just as the central angles are parts of the whole, each central angle intersects the circle to make an arc. For any circle, the arc measure is equal to the central angle measure in radians. If a circle has a radius that is not 1 unit, the arc length is proportional to the radius. If a circle has a radius of 5, for example, then the arc length is 5 times the central angle measure.

Practice

Find the indicated measures.

40. The radian measure of a central angle that is 120°

41. The central angle in degrees for the wedge that is 40% of a circle

42. The radian measure of an arc that is 25% of a circle

43. The arc length that is 60% of a circle with radius of 3 units

Practice Answers

1. acute
2. reflex
3. right
4. obtuse
5. straight
6. The complement is 28°; the supplement is 118°.
7. The complement is 72°; the supplement is 162°.
8. The complement is 6°; the supplement is 96°.
9. $m\angle x = 30°$, and $m\angle y = 30°$
10. 38°
11. $m = 30$; angles are 120° and 60°.
12. $m = 7$; the angle measure is 56°.
13. 40°
14. 140°
15. 40°
16. $p = 30$; the angle measure is 105°.
17. $p = 30$; the angle measure is 105°.
18. $p = 30$; the angle measure is 75°.
19. right isosceles
20. obtuse scalene
21. acute isosceles
22. 40°
23. 55°
24. 138°
25. 63°
26. 47°
27. 70°
28. true
29. true
30. false
31. false
32. true
33. true
34. $m\angle x = 135°$
35. $11x = 15x - 40$, so $x = 10$. The angles are 110°, 70°, 110°, 70°.
36. $m\angle w = 120°$; $m\angle x = 60°$; $m\angle y = 100°$; $m\angle z = 80°$

37. one

38. four

39. none

40. $\frac{2\pi}{3}$ radians

41. 144°

42. $\frac{\pi}{42}$ radians

43. 3.6π units, or approximately 11.3 units.

MEASUREMENT
For the COMPASS, ASSET, and ACCUPLACER

If I am given a formula, and I am ignorant of its meaning, it cannot teach me anything, but if I already know it what does the formula teach me?

—ST. AUGUSTINE

This chapter covers a lot of ground, and might take you longer than 20 minutes to complete. Split up your study into two or three parts if you need extra time to make your way through this lesson's examples.

LESSON SUMMARY

This lesson covers measurement, an important application of mathematics. It allows us to find the lengths of objects, calculate area of a region, and find the amount that can fill a three-dimensional object. This lesson will walk you through the steps to using the Pythagorean theorem, calculating perimeter, finding the area and circumference of circles, finding the area of polygons and irregular regions, and finding the surface area and volume of solids.

The Pythagorean Theorem

In a right triangle, the two sides that form the right angle are known as the legs, and the side across from the right angle is known as the hypotenuse. The sides are commonly labeled a, b, and c, where a and b are the legs and c is the hypotenuse, as shown in this figure.

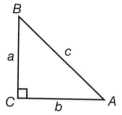

When studying the sides of a right triangle, an important relationship emerges. The sum of the squares of the two legs of the right triangle is equal to the square of the length of the hypotenuse.

QUICK FACT

The Pythagorean theorem can be summarized as $a^2 + b^2 = c^2$.

The Pythagorean theorem can be used to find a missing side of a right triangle. Take the following examples.

Example 1

The two legs of a right triangle measure 3 m and 4 m. What is the measure of the hypotenuse?

To solve for the missing side, substitute into the formula $a^2 + b^2 = c^2$ and use $a = 3$, $b = 4$.

Substitute:
$$3^2 + 4^2 = c^2$$
Apply the exponents:
$$9 + 16 = c^2$$
Add:
$$25 = c^2$$
Take the square root of each side of the equation:
$$\sqrt{25} = \sqrt{c^2}$$
$$5 = c$$
The hypotenuse is 5 units long.

Example 2

A leg of a right triangle measures 6 in. and the hypotenuse measures 12 in. Find the measure of the other leg of the right triangle.

To solve for the missing side, substitute into the formula $a^2 + b^2 = c^2$ and use $a = 6$, and $c = 12$.

Substitute:
$$6^2 + b^2 = 12^2$$
Apply the exponents:
$$36 + b^2 = 144$$
Subtract 36 from each side of the equation:
$$36 - 36 + b^2 = 144 - 36$$
$$b^2 = 108$$
Take the square root of each side of the equation:
$$\sqrt{b^2} = \sqrt{108}, \text{ so } b = \sqrt{108} \approx 10.392$$

Example 3

A flagpole is located 10 meters away from a park bench. If the direct distance from the park bench to the top of the pole is 26 meters, what is the height of the flagpole?

First, sketch a picture of the situation in order to identify each of the sides. A possible sketch could be as follows:

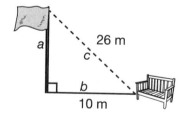

The flagpole forms a right angle with the ground, so the figure contains a right triangle. Since the height of the flagpole is a, the distance to the bench is b, and the direct distance between the bench and the top of the flagpole is c, set up the equation $a^2 + b^2 = c^2$ and use $b = 10$, and $c = 26$. Keep in mind that a and b

are interchangeable because of the commutative property of addition.

Substitute:
$$a^2 + 10^2 = 26^2$$

Apply the exponent:
$$a^2 + 100 = 676$$

Subtract 100 from each side of the equation:
$$a^2 - 100 = 676 - 100$$
$$a^2 = 576$$

Take the square root of each side of the equation:
$$\sqrt{a^2} = \sqrt{576}, \text{ so } a = 24$$

The height of the flagpole is 24 meters.

Practice

Use the Pythagorean theorem to find the missing side of each right triangle. Round to the nearest hundredth, if necessary.

1. $a = 6, b = 8, c = ?$

2. $a = 11, c = 14.1, b = ?$

3.

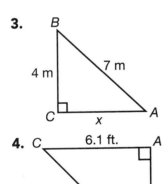

4.

Perimeter

The *perimeter* of a two-dimensional figure is the distance around the figure, or the sum of the length of the sides. To find the perimeter of any object, add the lengths of the sides around the outside of the figure.

Example 1
Find the perimeter of a square with side length 7 ft.

The formula for the perimeter of a square is $P = s + s + s + s$, or $P = 4s$.

Substitute into the formula to get $P = 4(7) = 28$. The perimeter is 28 ft.

Example 2
Find the perimeter of this triangle:

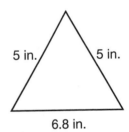

The formula for the perimeter of a triangle is $P = s_1 + s_2 + s_3$. Substitute into the formula to get $P = 5 + 5 + 6.8 = 16.8$. The perimeter is 16.8 in.

Example 3
Find the perimeter of a rectangle with a length of 10.5 m and a width of 12 m.

The formula for the perimeter of a rectangle is $P = l + w + l + w$, or $P = 2l + 2w$. Substitute into the formula to get $P = 2(10.5) + 2(12) = 21 + 24 = 45$. The perimeter is 45 m.

Example 4

Find the perimeter of a right triangle with legs of 6 m and 8 m.

In order to find the perimeter, you must first find the length of the hypotenuse. Use the Pythagorean theorem and the length of the two legs to find the length of this side:

$$a^2 + b^2 = c^2$$

Use $a = 6$, and $b = 8$ and substitute:

$$6^2 + 8^2 = c^2$$

Apply the exponent:

$$36 + 64 = c^2$$

Add:

$$100 = c^2$$

Take the square root of each side of the equation:

$$\sqrt{100} = \sqrt{c^2}, \text{ so } c = 10$$

Now, find the sum of the three sides to calculate the perimeter:

$$P = 6 + 8 + 10 = 24 \text{ m}$$

Practice

Find the perimeter of each of the following.

5. A square with side length of 12 in.

6. A triangle with sides of 11.2 m, 13.5 m, and 16 m.

7. A rectangle with a length of 16.5 ft. and a width of 18 ft.

8. A scalene right triangle with one leg that measures 8 units, and a hypotenuse of 17 units.

Area

When finding the area of a particular region, you are calculating the number of square units that cover the region, as in the following figure with an area of 12 square units. Be sure to label all areas with square units.

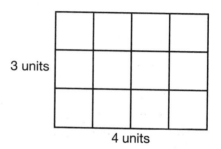

3 units

4 units

QUICK FACTS

The formula for the area of a parallelogram is $A = base \times height$, or $A = b \times h$.

The formula for the area of a triangle is $A = \frac{1}{2} \times base \times height$, or $A = \frac{1}{2} \times b \times h$.

The formula for the area of a trapezoid is $A = \frac{1}{2} \times height(base_1 + base_2)$, or $A = \frac{1}{2}h(b_1 + b_2)$.

Because all squares, rectangles, and rhombuses are also parallelograms, the areas of these figures can also be found by using the formula $A = base \times height$.

Example 1

Find the area of a rectangle with a base of 13 in. and a height of 11 in.

To find the area, substitute into the corresponding formula, $A = b \times h$:

$$A = 13 \times 11 = 143 \text{ in.}^2$$

Example 2

Find the area of a triangle with a base of 7 m and a height of 20 m.

To find the area, substitute into the corresponding formula, $A = \frac{1}{2} \times b \times h$:

$A = \frac{1}{2}(7 \times 20) = \frac{1}{2}(140) = 70$. The area is 70 m².

Example 3

Find the area of this trapezoid:

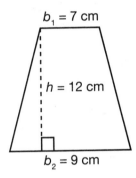

$b_1 = 7$ cm

$h = 12$ cm

$b_2 = 9$ cm

To find the area, substitute into the corresponding formula, $A = \frac{1}{2}h(b_1 + b_2)$:

$A = \frac{1}{2} \times 12(7 + 9) = \frac{1}{2} \times 12(16) = 96$. The area is 96 cm².

Example 4

If the area of a triangle is 24 square units and the base measures 4 units, what is the height of the triangle?

To find the height, substitute into the corresponding formula, $A = \frac{1}{2} \times b \times h$:

$24 = \frac{1}{2}(4 \times h)$

The equation simplifies to $24 = 2h$. Divide each side of the equation by 2 to get $12 = h$. The height of the triangle is 12 units.

Practice

For Questions 9 through 11, find the area of the figure.

9. A rectangle with a base of 14.1 meters and a height of 3.2 meters.

10. A trapezoid with one base of 10 ft., another base of 12 ft., and a height of 8 ft.

11. A triangle with a base of 10 cm and a height of 12.6 cm.

12. If the area of a parallelogram measures 68 in.² and the base is 17 in., what is the height of the figure?

Circles

Circumference of Circles

The *circumference* of a circle is the special term for the distance around the circle, and is similar to perimeter. The circumference either can be expressed in terms of π (which means the symbol appears in the answer), or can be approximated using a decimal value in place of π. Use the π button on a calculator or the decimal 3.14.

QUICK FACTS

The formula for circumference of a circle is $C = \pi d$, where d is the diameter of the circle. Recall that the diameter is twice the radius of any circle.

Example 1

The circumference of a circle with a diameter of 8 cm can be found by substituting $d = 8$ into the formula:

$C = \pi \times 8 = 8\pi$ cm, in terms of π

The approximate circumference can be found by substituting 3.14 for the symbol π:

$8\pi = 8(3.14) = 25.13$ cm

Example 2

The circumference of a circle with a radius of 15 mm can be found by first doubling the radius to find the diameter of 30 mm. Then, substitute $d = 30$ into the formula:

$C = \pi \times 30 = 30\pi$ mm, in terms of π

The approximate circumference can be found by substituting 3.14 for the symbol π:

$30\pi = 30(3.14) = 94.2$ mm

Area of Circles

The area of a circle is the number of square units that can fill the circle. The area can be expressed in terms of π, which means the symbol appears in the answer, or can also be approximated using a decimal approximation for π.

QUICK FACT

The formula for the area of a circle is $A = \pi r^2$, where r is the radius of the circle.

Example 1

The area of a circle with a radius of 6 cm can be found by substituting $r = 6$ into the formula $A = \pi r^2$. The formula becomes $A = \pi(6)^2 = 36\pi$ cm^2, in terms of π.

Example 2

The area of a circle with a diameter of 17 m can be found by first dividing the diameter by 2 to get 8.5 and substituting $r = 8.5$ into the formula $A = \pi r^2$. The formula becomes $A = \pi(8.5)^2 = 72.25\pi$ m^2, in terms of π.

The following example incorporates both the area and the circumference of a circle.

Example 3

If the area of a circle is equal to 25π square units, what is the circumference of the circle?

Because the area is equal to 25π units^2, then $r^2 = 25$, so $r = 5$. If the radius is equal to 5 units, then the diameter is equal to 10 units. The circumference is

$c = \pi d = 10\pi$ units.

Practice

For Questions 13 through 15, find the circumference and area of each circle in terms of π, given the radius or diameter.

13. $r = 14$ m

14. $d = 20$ cm

15. $r = 7.5$ in.

16. What is the area of a circle with a circumference of 26π units?

Perimeter and Area of Irregular Regions

Irregular regions are shapes that are made up of two or more other regions, typically different polygons. To find the perimeter of any irregular region, determine the length of each outer side of the shape and find the sum of these sides.

Example
Find the perimeter of this figure:

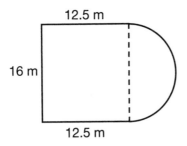

The figure is composed of a rectangle and a semicircle. The distance around the figure can be found by adding the three outer sides of the rectangle with half of the circumference of the circle with the same diameter. The perimeter then becomes $P = s_1 + s_2 + s_3 + \frac{1}{2}\pi d$. Because the figure on the right is only a semicircle, note that only half of the circumference is calculated. Substitute into the formula:

$$P = 12.5 + 12.5 + 16 + \frac{1}{2}\pi(16) = 41 + 8\pi$$
$$= 66.13 \text{ m, to the nearest hundredth}$$

To find the area of any irregular region, break the entire shape down into smaller regions, find the area of each region, and then add them together to find the sum of all the areas.

Example
The following figure is made up of a square and a right triangle.

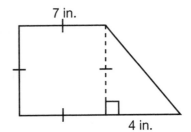

Find the area of each figure, and then add the areas together to find the total. The area is the area of the triangle added to the area of the square:

$$A = \frac{1}{2}bh + s^2 = \frac{1}{2}(4)(7) + (7)2 = \frac{1}{2}(28) + 49 = 14 + 49 = 63 \text{ in.}^2$$

Another type of irregular region is one that has a missing interior area.

Example

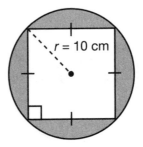

In this figure, there is a square inscribed in a circle with a radius of 10 cm. To find the area of the shaded region, find the area of the outer region (circle) and subtract the area of the inner region (square). Before this, use the Pythagorean theorem to find the length of the side of the square.

To solve for the missing side, substitute into the formula $a^2 + b^2 = c^2$ and use $a = b$, and $c = 20$. Since the radius is 10 cm, the diameter is 20 cm and forms the hypotenuse.

Substitute:

$$a^2 + a^2 = 20^2$$

Combine like terms and apply the exponent:

$$2a^2 = 400$$

Divide each side of the equation by 2:

$$a^2 = 200$$

Take the square root of each side of the equation:

$\sqrt{a^2} = \sqrt{200}$, so $a = \sqrt{200} \approx 14.14$ cm

So the area becomes $A = \pi r^2 - s^2 = \pi(10)^2 - (14.14)^2 \approx 100\pi - 199.9396 \approx 114.22$ cm².

Practice

Find each of the indicated measurements.

17. Find the perimeter of this figure:

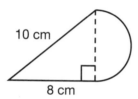

10 cm

8 cm

18. Find the perimeter of this figure:

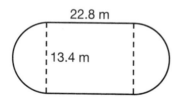

22.8 m

13.4 m

19. Find the area of the figure in Question 18.

20. Find the area of this figure:

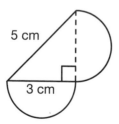

5 cm

3 cm

21. If the area of the square is 36 m², what is the perimeter?

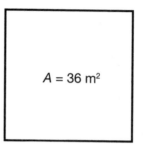

A = 36 m²

Surface Area

The surface area of a three-dimensional solid is exactly that—the total area of all of the solid's surfaces. To find the surface area, find the area of each surface, or face, and then find the sum of the areas of the faces.

Cube

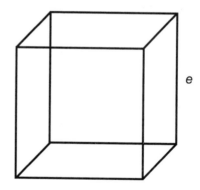

e

A cube has six congruent faces that are all squares, so the surface area is equal to $SA = 6e^2$, where e is the length of one side, or edge, of the cube.

Example 1

Find the surface area of a cube with an edge length of 7 m.

Substitute into the formula:

$$SA = 6e^2 = 6(7)^2 = 6(49) = 294 \text{ m}^2$$

In a problem where the surface area is known, work backward to find the measure of the edge of the cube.

Example 2

If the surface area of a cube is 150 units^2, what is the length of an edge of the cube?

Substitute into the formula:

$$SA = 6e^2 \text{ becomes } 150 = 6e^2$$

Divide each side of the equation by 6:

$$\frac{150}{6} = \frac{6e^2}{6} \text{ becomes } 25 = e^2$$

Take the square root of each side of the equation to get $5 = e$. The length of an edge of the cube is 5 units.

Rectangular Prism

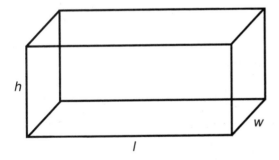

In a rectangular prism, the top area is the same as the bottom, the left side is the same as the right side, and the front is the same as the back. Thus, the formula for the surface area can be summarized as $SA = 2lw + 2lh + 2hw$.

Example

What is the surface area of a box in the shape of a rectangular prism with a height of 60 in., a length of 12 in., and a width of 42 in.?

Substitute into the formula:

$$SA = 2lw + 2lh + 2hw \text{ becomes } SA = 2(12)(42) + 2(12)(60) + 2(60)(42)$$

Multiply:

$$SA = 1{,}008 + 1{,}440 + 5{,}040$$

Add to get the surface area of the box:

$$SA = 7{,}488 \text{ in.}^2$$

Cylinder

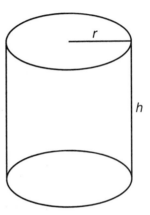

The bases of any cylinder are circles, so the surface area is the area of the two circles plus the area of the side of the cylinder. This can be summarized as $SA = 2\pi r^2 + \pi dh$, where r is the radius of the circular bases, d is the diameter of the bases, and h is the height of the cylinder. The πdh part of the equation comes from the fact that the cylinder can be unrolled into a rectangle of length h and width πd.

Example

A can in the shape of a cylinder has a height of 6 inches and a base with a radius of 2 inches. What is the surface area of the can to the nearest hundredth?

Substitute into the formula:

$$SA = 2\pi r^2 + \pi dh \text{ becomes } SA = 2\pi(2)^2 + \pi(4)(6)$$

Apply the exponent and multiply:

$$SA = 2\pi(4) + \pi(4)(6) = 8\pi + 24\pi$$

Add:

$$SA = 32\pi \approx 100.53 \text{ in.}^2$$

Sphere

The surface area of a sphere can be summarized as the area of four circles with the same radius, so the surface area formula is $SA = 4\pi r^2$.

Example
Find the surface area to the nearest hundredth for a basketball with a radius of 4.5 inches.
Substitute into the formula:
$$SA = 4\pi r^2 \text{ becomes } SA = 4\pi(4.5)^2$$
Evaluate the exponent and multiply:
$$SA = 4\pi(20.25) = 81\pi \approx 254.47 \text{ in.}^2$$

TIP

Remember that surface area, like general area, is always labeled in square units.

Practice
For Questions 22 through 25, find the surface area of each of the following with the given dimensions.

22. A cube with an edge of 2 cm

23. A rectangular prism with a length of 12 m, a height of 30 m, and a width of 5.8 m

24. A cylinder with a height of 17.3 in. and a base radius of 14 in.

25. A sphere with a radius of 6.2 mm

26. How much more surface area does a cube with an edge length of 16 m have than a rectangular prism with a height of 10 m, a width of 15 m, and a length of 18 m?

27. What is the edge length of a cube with a surface area of 864 cm²?

Volume

The *volume* of a three-dimensional solid is the number of unit cubes that can fit inside a solid. Volume is expressed in cubic units.

QUICK FACT

The formula for the volume for most three-dimensional solids can be summarized as $V = Bh$, where B = the area of the base, and h = the height of the solid. If the solid comes to a point at one base, the formula is $V = \frac{1}{3}Bh$, where B = the area of the base, and h = the height of the solid. This type of solid is always one-third of the volume of the same solid with equal bases instead of the point.

TIP

Because volume is the amount of space a three-dimensional object occupies, volume is always labeled in cubic units.

The formulas for the following solids are based on the Quick Fact.

Cube

$V = e^3$, where e is the length of a side, or edge, of the cube.

Example

Find the volume of a cube with an edge length of 8 cm.

Substitute into the formula and apply the exponent:

$$V = e^3 \text{ becomes } V = (8)^3 = 512 \text{ cm}^3$$

Rectangular Prism

$V = lwh$, where l is the length, w is the width, and h is the height of the prism.

Example

If the volume of a rectangular prism is 243 m³, what is the length if the width is 9 m and the height is 6 m?

Substitute into the formula:

$$V = lwh \text{ becomes } 243 = (l)(9)(6)$$

Multiply to get $243 = 54l$.

Divide each side of the equation by 54 to get $l = 4.5$. The length is 4.5 m.

Triangular Prism

$V = Bh$, where B is the area of the triangular base, and h is the height of the prism.

Example

Find the volume of a triangular prism with a base area of 50 cm² and a height of 1.75 cm.

Substitute into the formula and multiply:

$$V = Bh \text{ becomes } V = (50)(1.75) = 87.5 \text{ cm}^3$$

Cylinder

$V = \pi r^2 h$, where r is the radius of the circular base and h is the height of the cylinder.

Example

Find the height of a cylinder with a radius of 15 mm and volume of 2,475π mm³.

Substitute into the formula:

$$V = \pi r^2 h \text{ becomes } 2{,}475\pi = \pi(15)^2 h$$

Apply the exponent ($15^2 = 225$) and divide each side by 225π:

$$\frac{2{,}475\pi}{225\pi} = \frac{225\pi h}{225\pi}$$

$h = 11$, so the height of the cylinder is 11 mm.

Cone

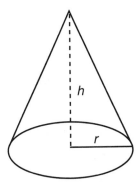

$V = \frac{1}{3}\pi r^2 h$, where r is the radius of the circular base and h is the height of the cone.

Example

A cone has a radius of 2.5 units and a height of 16 units. What is the volume to the nearest hundredth?

Substitute into the formula:

$$V = \frac{1}{3}\pi r^2 h \text{ becomes}$$
$$V = \frac{1}{3}\pi(2.5)^2(16)$$

Apply the exponent and multiply:

$$V = \frac{1}{3}(100)\pi = \frac{100\pi}{3} \approx 104.72 \text{ units}^3$$

Pyramid

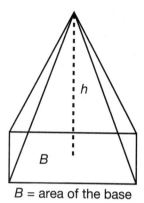

B = area of the base

$V = \frac{1}{3}Bh$, where B is the area of the base, and h is the height of the pyramid.

Example

A pyramid has a base area of 80 cm² and a height of 70 cm. What is the volume of the pyramid to the nearest hundredth?

Substitute into the formula:
$$V = \tfrac{1}{3}Bh \text{ becomes } V = \tfrac{1}{3}(80)(70)$$
Multiply and then divide by 3:
$$V = \tfrac{1}{3}(5,600) = \tfrac{5,600}{3} = 1,866.\overline{6} \approx$$
1,866.67 cm³

Sphere

$V = \frac{4}{3}\pi r^3$, where r is the radius of the sphere.

Example

What is the volume of a sphere with radius 7 mm, to the nearest hundredth?

Substitute into the formula:
$$V = \tfrac{4}{3}\pi r^3 \text{ becomes } V = \tfrac{4}{3}(7)^3$$
Apply the exponent and multiply:
$$V = \tfrac{4}{3}\pi(343) = \tfrac{1,372\pi}{3} \approx 1,436.03 \text{ mm}^3$$

Practice

For each problem, find the volume or indicated dimension.

28. A cube has an edge length of 0.6 m. What is the volume of the cube?

29. A triangular prism has a base area of 45 cm² and a height of 15.3 cm. What is the volume of the prism?

30. A file cabinet in the shape of a rectangular prism has a height of 5 ft., a width of 1.5 ft., and a total volume of 30 ft.³. What is the length of the cabinet?

31. A soup can in the shape of a cylinder has a base radius of 8 cm and a height of 15 cm. What is the volume of the can, in terms of π?

32. A cone has a volume of 12π. If the base radius is 3 units, what is the height of the cone?

33. A pyramid has a height of 6.6 m and a volume of 11.66 m³. What is the area of the base of the pyramid?

34. If a sphere has a radius of 16 cm, what is the volume of the sphere to the nearest hundredth?

35. A ball in the shape of a sphere has a volume of 307.55 mm³. What is the length of the radius of the ball?

Practice Answers

1. $c = 10$
2. $b \approx 8.82$
3. $x \approx 5.74$ m
4. $x \approx 8.63$ ft.
5. 48 in.
6. 40.7 m
7. 69 ft.
8. $8 + 15 + 17 = 40$
9. 45.12 m^2
10. 88 ft.^{2}
11. 63 cm^2
12. 4 in.
13. $C = 28\pi$ m; $A = 196\pi$ m^2
14. $C = 20\pi$ cm; $A = 100\pi$ cm^2
15. $C = 15\pi$ in.; $A = 56.25\pi$ in.^{2}
16. 169π units^2
17. 27.42 cm
18. 87.70 m
19. 446.55 m^2
20. 15.82 cm^2
21. 24 m
22. 24 cm^2
23. 1,207.2 m^2
24. 2,753.29 in.^{2}
25. 483.05 mm^2
26. The cube is larger by 336 m^2.
27. 12 cm
28. 0.216 m^3
29. 688.5 cm^3
30. 4 ft.
31. 960π cm^3
32. 4 units
33. 5.3 m^2
34. 17,157.28 cm^3
35. 4.19 mm

12 ▶ COORDINATE GEOMETRY For the COMPASS, ASSET, and ACCUPLACER

It is not enough to have a good mind; the main thing is to use it well.

—RENÉ DESCARTES

LESSON SUMMARY

You reviewed plane geometry in Lesson 10. René Descartes was a philosopher and mathematician in the 1600s who is credited with the discovery of the connection between algebra and geometry. The Cartesian coordinate system is named after him. This lesson reviews how to plot points on this coordinate grid system, how to graph equations and inequalities, and how to calculate significant measures such as the distance between two points and the slope of a linear equation.

The Cartesian Coordinate System

The coordinate system is the intersection of two number lines, called axes. One is horizontal and the other is vertical. The point where the two axes intersect is called the origin, and is given the coordinates of (0,0).

The Cartesian coordinate system is used to locate points and other two-dimensional geometric figures. A point has two coordinates, also called an ordered pair, (x,y), which indicate how far the point is from the origin. The x-coordinate indicates how far a point is from the origin in the left/right direction. Points to the right of the origin have a positive *x*-coordinate; points to the left of the origin have a negative *x*-coordinate. The *y*-coordinate indicates how far a point is from the origin in the up/down direction. Points above the origin have a positive *y*-coordinate; points below the origin have a negative *y*-coordinate.

Here is an example of a coordinate plane:

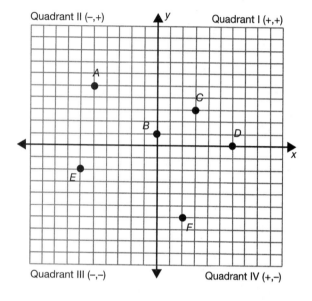

The coordinate plane is divided into four equal quadrants, with the origin in the center. Quadrant I is in the upper right corner, and the quadrants are numbered in a counterclockwise direction. The figure shows the sign of the ordered pairs based on which quadrant they lie in. The coordinates of the labeled points are: A (−5,5), B (0,1), C (3,3), D (6,0), E (−6,−2), and F (2,−6). The point (0,0) is the origin, and is not in any of the quadrants. Likewise, any point on either the *x*-axis or the *y*-axis is not in any quadrant, but is said to be on the respective axis.

Practice

1. Name the points on the coordinate plane:

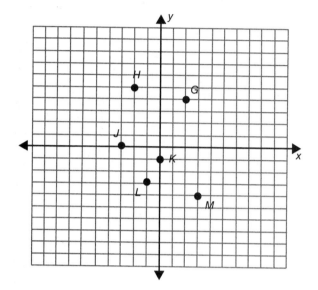

For Questions 2 through 6, plot the points on the coordinate plane:

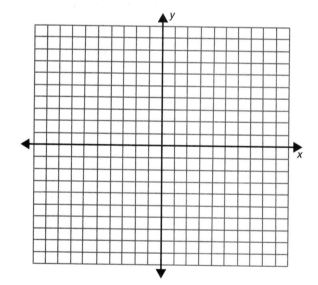

2. N (4,2)

3. P (0,−3)

4. $Q\,(-1,7)$

5. $R\,(-5,-2)$

6. $S\,(3,-1)$

The Midpoint

DEFINITION

The *midpoint* of the line segment with endpoints given by (x_1,y_1) and (x_2,y_2). The formula is midpoint $= (\frac{x_1 + x_2}{2}, \frac{y_1 + y_2}{2})$.

Example

Find the midpoint of the line segment with the endpoints $(-5,2)$ and $(9,-4)$.

$$(\tfrac{-5 + 9}{2}, \tfrac{2 + (-4)}{2}) = (\tfrac{4}{2}, \tfrac{-2}{2}) = (2,-1)$$

You can find the center of a circle given two endpoints of a diameter to the circle.

Example

If the endpoints of a diameter to a circle are $(6,10)$ and $(-10,8)$, then the center of the circle is:

$$(\tfrac{6 + (-10)}{2}, \tfrac{10 + 8}{2}) = (-2,9)$$

The Distance between Two Points

In Lesson 11, you reviewed the Pythagorean theorem. If you look at the following coordinate plane, you can see that each of the dotted line segments between two points is the hypotenuse of a right triangle. The distance formula is a form of the Pythagorean theorem. Since $a^2 + b^2 = c^2$, you can take the square root of both sides of this equation to get the distance $c =$

$\sqrt{a^2 + b^2}$. The value a can be considered the horizontal distance and b is the vertical distance.

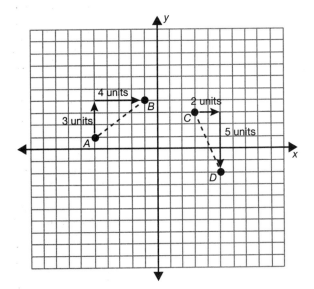

Example

So, for example, to find the distance between points A and B use the Pythagorean theorem to get $c^2 = 4^2 + 3^2$, or $c^2 = 9 + 16$. Thus $c = \sqrt{25} = 5$. Another example is to again use the theorem to find the distance between C and D: $c^2 = 2^2 + 5^2$, or $c^2 = 4 + 25$. Therefore $c = \sqrt{29}$, or 5.39 units.

DEFINITION

The distance, d, between two points on the coordinate plane (x_1,y_1) and (x_2,y_2) is $d = \sqrt{(x_2 - x_1)^2 + (y_2 - y_1)^2}$.

In this coordinate plane, the distance between $A\,(-5,1)$ and $B\,(-1,4)$ was shown to be 5 units. Using the distance formula, the distance is $d = \sqrt{(-1 - -5)^2 + (4 - 1)^2} = \sqrt{(4)^2 + (3)^2} = \sqrt{25} = 5$. The distance between $C\,(3,3)$ and $D\,(5,-2)$ is $d = \sqrt{(5 - 3)^2 + (-2 - 3)^2} = \sqrt{(2)^2 + (-5)^2} = \sqrt{29}$, or approximately 5.39 units.

Practice

Given the endpoints of a segment, find the midpoint.

7. (−14,3) and (8,−6)

8. (2,−7) and (5,2)

9. Find the center of the circle that has (6,−9) and (6,3) as endpoints of a diameter.

Find the distance between the given points.

10. A (2,4) and B (−3,16)

11. C (1,−1) and D (7,7)

12. E (−2,−3) and F (6,−4)

Slope

In addition to the distance between two points, there is another measure associated with two points, called slope. Like the distance formula, you calculate slope using the point coordinates.

Slope is often referred to as "rise over run," meaning the vertical distance over the horizontal distance between the points. In the preceding figure, the slope of segment AB is $\frac{3}{4}$, because you go up 3 and right 4 to get from A to B (both in the positive direction), or down 3, left 4 to get from B to A. The slope of segment CD is $\frac{-5}{2}$, because you go down 5 and right 2 from C to D, or up 5 and left 2 to get from D to C (one in the positive direction and one in the negative direction).

Practice

Use this figure to answer Questions 13 through 16.

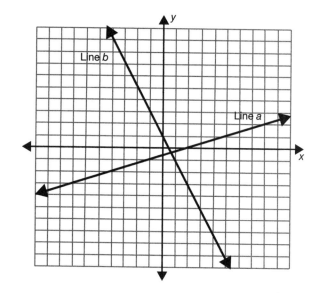

13. Which line has the larger slope?

14. Which line is steeper?

15. Which line has a negative slope?

16. Which line has a positive slope?

17. Find the slope between the points E (–3,4) and F (–1,–6).

18. Find the slope between the points G (5,8) and H (–1,5).

Graphing Equations on the Coordinate Plane

Linear equations demonstrate a relationship between two variables. Because there are two variables, the solution set can be shown on the two-dimensional coordinate plane. Because it is a linear equation, the graph is a straight line.

Linear equations have a slope-intercept form of $y = mx + b$, where m represents the slope and b is a constant that represents the y-intercept, or the y-coordinate when $x = 0$. On the graph where $x = 0$, the point is on the y-axis.

To graph a linear equation, plot the y-intercept. To get other points, convert the slope value to a fraction, $\frac{rise}{run}$, and move according to the fraction in the vertical direction (the rise) and in the horizontal direction (the run). Repeat this method to get a third point. Then connect the three points to form the line that is the solution set to the equation. The following graph shows a line that corresponds with the linear equation $y = 2x + 3$.

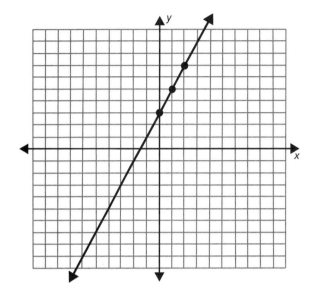

To determine the linear equation based on a given graph, such as the one given next, find the y-intercept, b, and determine the slope, m, as reviewed earlier. In the graph, the y-coordinate where the line crosses the y-axis is (0,–2). To get from that point to the other point shown you would travel up 4 units and to the left 1 unit to get a slope of $\frac{4}{-1} = -4$. Therefore, the equation is $y = -4x - 2$.

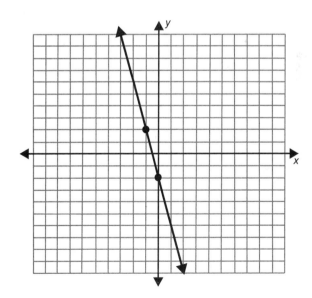

Special Linear Equations

You should be familiar with two special linear equations—that of a horizontal line and that of a vertical line. Refer to the following figure.

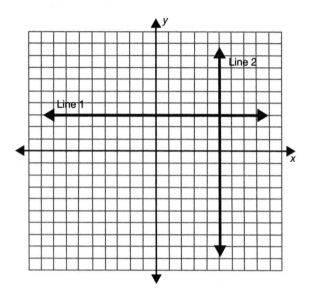

A horizontal line has a slope of zero. Whatever the value of x, the value of y is constant. For example, line 1 in the figure has points of $(4,3)$ and $(-6,3)$; the slope is therefore $\frac{3-3}{-6-4} = \frac{0}{-10} = 0$. Since $m = 0$, the equation for this horizontal line is $y = 3$.

A vertical line has a slope that is undefined. In a vertical line, whatever the value of y, the value of x is constant. Line 2 in the figure has points $(5,10)$ and $(5,-7)$, and the slope is therefore $\frac{-7-10}{5-5} = \frac{-17}{0}$. Any fraction with a denominator of 0 is undefined, and therefore the slope is said to be undefined. The equation of this vertical line is $x = 5$.

Practice

Determine the equation of the lines graphed.

19.

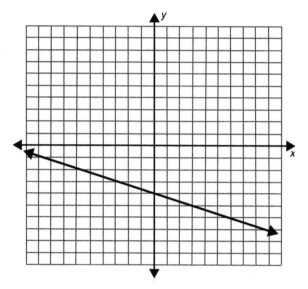

20.

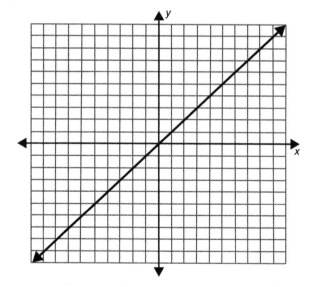

21.

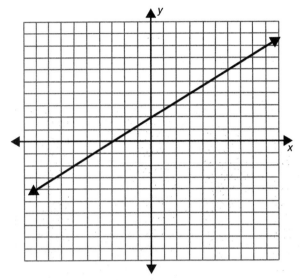

22.

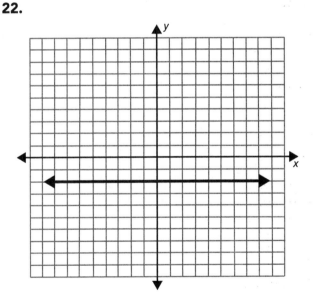

number of solutions and a ray on a number line was used to show the solution set. The solution set to an inequality in two variables also has an infinite number of solutions. The graph of this set is an area rather than a line.

To graph these inequalities, first graph the line as reviewed earlier, using the slope and y-intercept. If the symbol is ≤ or ≥, the borderline will be solid to show that points on the line are part of the solution set. If the symbol is < or >, the borderline will be dashed, since points on the borderline are not part of the solution set. The coordinate plane is then shaded based on the graphed borderline and whether the symbol is greater than or less than. If the inequality is > or ≥, shade the area above the line. If the inequality is < or ≤, shade the area below the line. The next plane shows the graph of the inequality $y > -2x$.

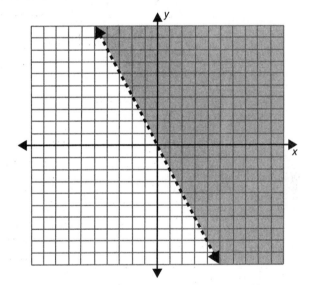

The y-intercept is 0, so the borderline passes through (0,0). The slope is $-2 = \frac{-2}{1}$, so from the origin you can travel up 2 and left 1, or down 2 and right 1. Other points are (−1,2) and (1,−2). The borderline is dashed because the symbol is simply >, and the area is shaded above because it is greater than.

Graphing Linear Inequalities on the Coordinate Plane

In Lesson 9, you reviewed how to solve and graph inequalities with one variable. There was an infinite

The next plane shows the graph of $y \leq \frac{1}{3}x + 2$.

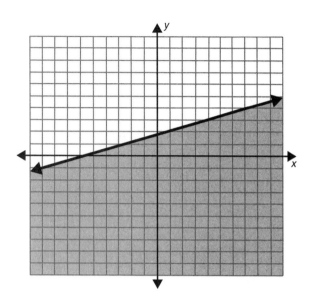

23.

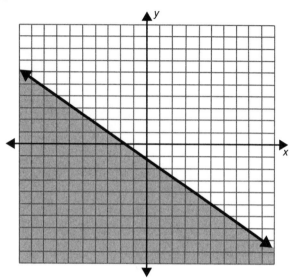

The *y*-intercept is 2, so the borderline passes through (0,2). The slope is $\frac{1}{3}$, so from the *y*-intercept you can travel up 1 and right 3, or down 1 and left 3. Other points are (−3,1) and (3,3). The borderline is solid because the symbol is ≤, and the area is shaded below because it is less than.

Practice

Name the inequality shown on the coordinate planes.

24.

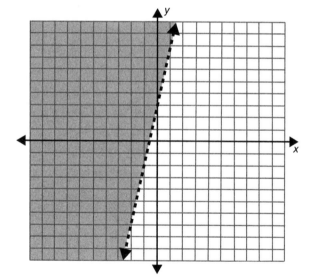

25.

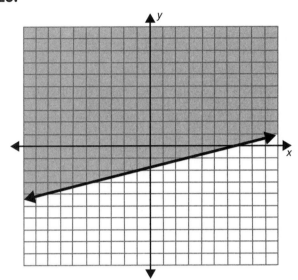

26.

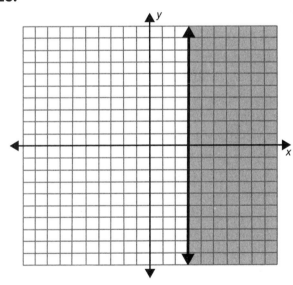

Practice Answers

1. G (2,4)
  H (−2,5)
  J (−3,0)
  K (0,−1)
  L (−1,−3)
  M (3,−4)

2–6.

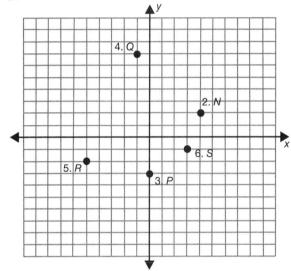

7. (−3,−1.5)
8. (3.5,−2.5)
9. (6,−3)
10. 13 units
11. 10 units
12. $\sqrt{65} \approx 8.06$ units
13. line a
14. line b
15. line b
16. line a
17. $\frac{-10}{2} = -5$
18. $\frac{3}{6} = \frac{1}{2}$
19. $y = -\frac{1}{3}x - 4$
20. $y = x$

21. $y = \frac{2}{3}x + 2$

22. $y = -2$

23. $y \leq \frac{3}{4}x - 1$

24. $y > 5x + 3$

25. $y \geq \frac{1}{4}x - 2$

26. $x \geq 3$

13 ▶ SYSTEMS OF EQUATIONS For the COMPASS, ASSET, and ACCUPLACER

Film is one of the three universal languages; the other two: mathematics and music.

—FRANK CAPRA

LESSON SUMMARY

In Lesson 12, you reviewed coordinate geometry and graphing linear equations and inequalities. In this lesson, you will brush up on how to solve simultaneous systems of equations and inequalities. A system of equations is a set of two or more equations that may or may not have a solution. This review covers two equations in two variables.

Simultaneous Linear Equations

A system of linear equations is two or more equations in two variables that are solved together. The solution is where the two graphed lines meet (intersect) on the coordinate plane, so the solution is most often an ordered pair. In the preceding lesson, you graphed a linear equation using the slope-intercept form. The graphed line showed all of the points, an infinite number, that would make the equation true. For example, look at this graph:

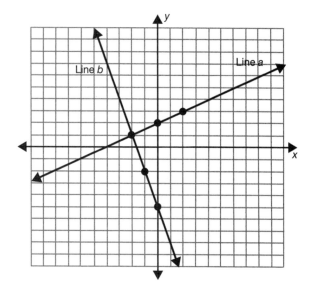

Line a, $y = \frac{1}{2}x + 2$, has points at (0,2), (−2,1), and (2,3). All of these ordered pairs, and the infinite other points that lie on the line, are solutions to the equation, by substituting in the values of x and y. The other line, $y = -3x - 5$, has its own set of infinite points that are solutions. One point, (−2,1), is a solution to both equations. This is the solution to the simultaneous set. You can verify this to be true.

Write the first equation:

$$y = \frac{1}{2}x + 2$$

Substitute in the ordered pair (−2,1) for x and y:

$$1 = \frac{1}{2}(-2) + 2$$

Simplify:

$$1 = -1 + 2$$
$$1 = 1 \qquad \text{Verified to be correct for this equation.}$$

Write the second equation:

$$y = -3x - 5$$

Substitute in the ordered pair (−2,1) for x and y:

$$1 = -3(-2) - 5$$

Simplify:

$$1 = 6 - 5$$
$$1 = 1 \qquad \text{Verified to be correct for this equation.}$$

You can find the solution to a system of equations by graphing the lines and finding the point where they intersect. You reviewed how to graph linear equations in the last lesson. By graphing both lines and finding their intersection, you can find the ordered pair that satisfies both equations.

QUICK FACTS

For simultaneous systems of equations, there may be one solution, no solutions, or an infinite number of solutions. If the lines are parallel, there is no solution. If the lines are the same line and fall right on top of each other on a graph, there is an infinite number of solutions.

Practice

Find the solution to the system of equations using the following graphs:

1.

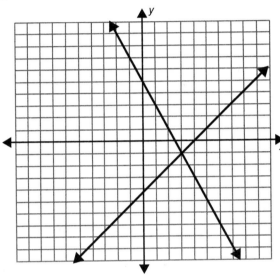

2.

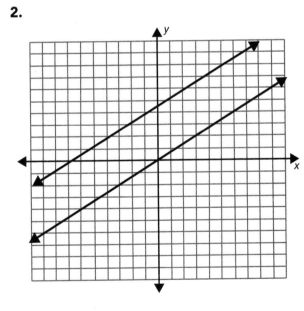

3.

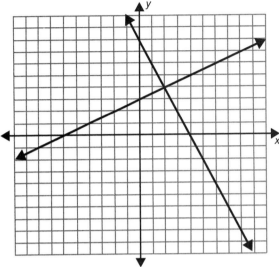

4.

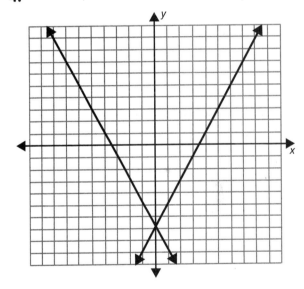

Parallel and Perpendicular Lines

Look at this graph. The two lines are parallel.

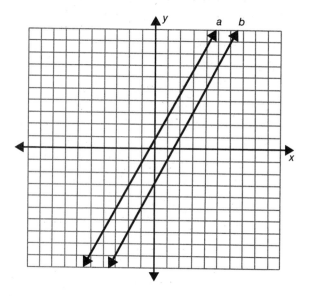

The equation for line a is $y = 2x + 1$ and for line b is given by $y = 2x - 3$. Notice that both equations have the same slope. This is always true of parallel lines, and it is a means for determining whether two lines are parallel by the equations alone.

Now look at this graph. These two lines are perpendicular.

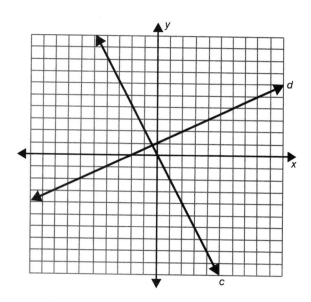

The equation for line c is $y = -2x$ and the equation for line d is $y = \frac{1}{2}x + 1$. Notice that the slopes of these lines are opposite reciprocals of each other; that is, one is positive and the other is negative, and the numerator and denominator are reversed. If you multiply the slopes, their product is -1. This fact always holds true for perpendicular lines, and can be used to identify perpendicularity. In order to determine this you may need to solve the equation for y to get the equations in slope-intercept form.

Example

For example, to determine the relationship of $y = -4x + 4$ and $4y - x = 8$, solve the second equation for y:

$4y - x = 8$ Write the equation.

$\underline{\;\; + x \;\; + x\;\;}$ Add x to both sides.

$\frac{4y}{4} = \frac{1}{4}x + 2$ Divide both sides (all terms) by 4.

$y = \frac{1}{4}x + 2$ Determine that the slope of this line is $\frac{1}{4}$.

The first equation has a slope of -4 and the second equation has a slope of $\frac{1}{4}$. These are negative reciprocals of each other, so the lines are perpendicular.

QUICK FACTS

Parallel lines have the same slope.
Perpendicular lines have slopes that are negative reciprocals of each other.

Practice

Determine whether the equations of these sets of lines are parallel, perpendicular, or neither.

5. $y = -5x + 1$ and $y = \frac{1}{5}x - 2$

6. $y = 4x$ and $y = \frac{1}{4}x + 6$

7. $y = 7x - 5$ and $y = 7x + 9$

8. $y = -\frac{1}{3}x$ and $6y + 2x = 12$

Graphic Solution to a System of Inequalities

In Lesson 12, you were reminded how to graph linear inequalities on the coordinate plane. Just like with equations, you can solve a system of linear inequalities. Instead of a point being the solution to this type of system, generally the solution to the system of inequalities is an area of a graph, where the graphs of the two inequalities overlap. Any point in this area should satisfy both of the inequalities. Recall that sometimes the borderlines are dashed, in which case those points on that border are not part of the solution. For example, the next figure shows the graph of the two inequalities $y \leq \frac{1}{2}x$ and $y > -2x + 1$. The region with the darkest shading is the solution to the system.

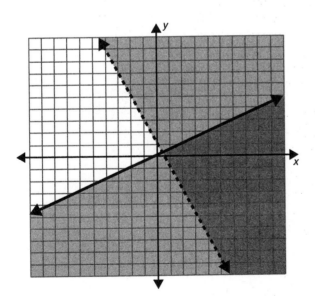

Sometimes a system of inequalities has no solution, as shown in the figure of the system $y \geq 3x$ and $y < 3x - 5$:

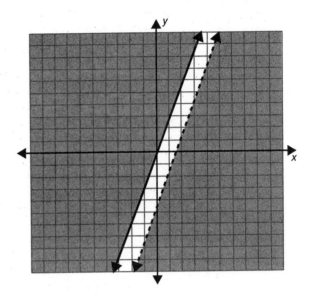

Practice

Name the two inequalities shown in the following systems. Then name a point in the solution.

9.

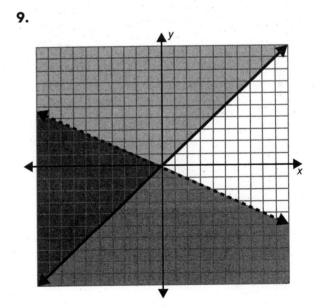

10.

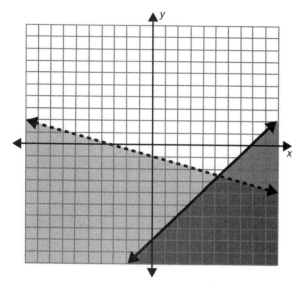

11.

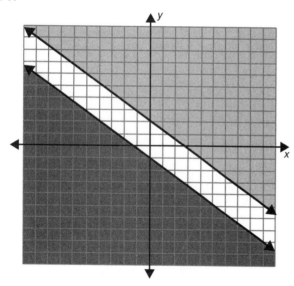

Solving Systems of Equations Algebraically

Sometimes the solution to a system of linear equations does not fall on the grid lines, and you cannot determine the exact solution by inspection of a graph. You can determine the solution to a system of equations algebraically by two different methods—substitution or elimination.

Solving Systems of Equations by Substitution

To solve a system of equations, you can use one of the equations to substitute into the other equation in place of one of the variables.

Example

To find the solution to the equations $y = 3x + 2$ and $y - 22 = -5x$, replace the y in the second equation with the equivalent expression $3x + 2$. This way, you will have one equation in one variable:

$$\begin{aligned} y - 22 &= -5x & &\text{Write the second equation.} \\ 3x + 2 - 22 &= -5x & &\text{Substitute in for } y \text{ from the} \\ & & &\text{first equation.} \\ 3x - 20 &= -5x & &\text{Combine the constant terms.} \\ \underline{-3x} &= \underline{-3x} & &\text{Subtract } 3x \text{ from both sides.} \\ -20 &= -8x & &\text{Simplify.} \\ \tfrac{-20}{8} &= \tfrac{-8x}{-8} & &\text{Divide both sides by } -8 \text{ to} \\ & & &\text{isolate } x. \\ 2.5 &= x \end{aligned}$$

The x-coordinate of the solution to the system is 2.5. Now use either one of the original equations to find the y-coordinate. Use the first equation and substitute in for x to get $y = 3(2.5) + 2$. Simplify to get $y = 7.5 + 2 = 9.5$. The solution is (2.5,9.5). You can verify that this is correct by checking it in the second equation. Check to be sure that $9.5 = -5(2.5) + 22$. It is true that 9.5 does in fact equal $-12.5 + 22$.

Practice

Find the solution to each system of equations. Use the algebraic substitution method.

12. $y = x - 8$ and $y = 4x + 16$

13. $y = 3(x + 2)$ and $y = 5x - 11$

14. $y = -2x + 4$ and $5y + 12x = -10$

Solving Systems of Equations by Elimination

Another method to solve a system of equations is to line up the equations and combine like terms in order to eliminate one of the variables. In order to do this, the equations must be in the same form, so that the variables will line up properly.

Example 1

To solve the system $-3x + 4y = 25$ and $3x - 2y = 1$, line up the equations:

$$\begin{array}{rl} -3x + 4y &= 25 \\ 3x - 2y &= 1 \\ \hline 2y &= 26 \end{array}$$ Add the equations; combine like terms.

$\frac{2y}{2} = \frac{26}{2}$ Divide both sides by 2.

$y = 13$ Simplify.

To find the value of x, substitute 13 for y in one of the equations:

$-3x + 4(13) = 25$

$-3x + 52 = 25$ Multiply.

$\underline{\quad -52 = -52}$ Subtract 52 from each side.

$-3x = -27$ Simplify.

$\frac{-3x}{-3} = \frac{-27}{-3}$ Divide both sides by -3.

$x = 9$

The solution is $(9,13)$.

Sometimes, it is necessary to multiply each term of one or both of the equations in order to eliminate one of the variables.

Example 2

To use the elimination method on the equations $2x - 5y = 15$ and $4x - 3y = 16$, multiply the first equation by -2 to get $-4x + 10y = -30$ and use the elimination method:

$$\begin{array}{rl} -4x + 10y &= -30 \\ 4x - 3y &= 16 \\ \hline 7y &= -14 \end{array}$$ Add the equations; combine like terms.

$\frac{7y}{7} = \frac{-14}{7}$ Divide both sides by 7.

$y = -2$ Simplify.

Substitute -2 for y to find the value of x in the first equation.

$2x - 5(-2) = 15$

$2x + 10 = 15$ Multiply.

$\underline{\quad -10 = -10}$ Subtract 10 from each side.

$2x = 5$ Simplify.

$\frac{2x}{2} = \frac{5}{2}$ Divide each side of the equation by 2.

$x = 2.5$

The solution is $(2.5,-2)$.

You may at times have to multiply both equations by a constant to eliminate a variable.

Example 3

Consider the example of solving the system of $2x + 3y = 5$ and $5x - 2y = 41$. You can choose to eliminate either variable; in this example, the variable x will be eliminated. This will require the first equation to be multiplied by 5 and the second equation by -2:

$5(2x + 3y = 5)$ Multiply the first equation through by 5.

$-2(5x - 2y = 41)$ Multiply the second equation through by -2.

$$\begin{array}{rl} 10x + 15y &= 25 \\ -10x + 4y &= -82 \\ \hline 19y &= -57 \end{array}$$ Distribute the 5. Distribute the -2. Add the equations; combine like terms.

$\frac{19y}{19} = \frac{-57}{19}$ Divide both sides by 19.

$y = -3$ The y-coordinate is negative 3.

By substituting $y = -3$ into the first equation, $x = 7$, so the solution is $(7,-3)$.

Systems That Have No Solution

When a system has no solution and you attempt to solve the system algebraically, the result will be a false statement.

Example

To solve the system $y = 4x - 3$ and $10 = 2y - 8x$, you can substitute $4x - 3$ from the first equation for y in the second equation:

$10 = 2(4x - 3) - 8x$	Substitute.
$10 = 8x - 6 - 8x$	Distribute the 2.
$10 \neq -6$	Combine like terms, and a false statement results.

This indicates that there is no solution, since no values of x and y will make $10 = -6$.

Practice

Find the solution to each system of equations. Use the algebraic elimination method.

15. $7x - 3y = 24$ and $3x + 3y = 6$

16. $3x + 2y = 15$ and $-x + y = 10$

17. $5x + 3y = 8$ and $x - y = -24$

18. $-2x + 4y = 12$ and $-\frac{1}{2}x + y = -4$

Practice Answers

1. $(3,-1)$
2. no solution (the empty set)
3. $(2,4)$
4. $(0,-7)$
5. perpendicular
6. neither
7. parallel
8. parallel
9. $y \geq x$ and $y < \frac{1}{2}x$; there are many answers to the solution, one of which is $(-6,-1)$.
10. $y \leq x - 8$ and $y < -\frac{1}{3}x - 1$; there are many answers to the solution, one of which is $(5,-5)$.
11. $y \geq -\frac{3}{4}x + 2$ and $y \leq -\frac{3}{4}x - 1$ (there is no solution—the empty set)
12. $(-8,-16)$
13. $(8.5,31.5)$
14. $(-15,34)$
15. $(3,-1)$
16. $(-1,9)$
17. $(-8,16)$
18. $0 \neq 28$, so there is no solution.

14 ▶ POLYNOMIALS AND FACTORING
For the COMPASS, ASSET, and ACCUPLACER

The essence of mathematics is not to make simple things complicated, but to make complicated things simple.

—S. GUDDER

This chapter covers a lot of ground, and might take you longer than 20 minutes to complete. Split up your study into two parts if you need extra time to make your way through this lesson's examples.

LESSON SUMMARY

Mathematics, and in particular algebra, can appear complicated at first. However, algebra by definition is generalized arithmetic, which means you are working with the basic operations of addition, subtraction, multiplication, and division, which you are already familiar with. Relating the procedures of algebra back to the basics of arithmetic will help unravel the mystery of algebra. This lesson reviews each of the types of polynomials, as well as the basic operations of algebraic expressions containing one or more terms. The second part of the lesson covers how to factor polynomials to help later in the simplifying and solving processes. At the conclusion, the lesson covers polynomial graphs.

To begin this lesson, first recall the following important information from Lesson 7.

DEFINITION

An *algebraic expression* is an expression that contains numbers and/or variables and at least one operation.

As stated in previous lessons, the number in front of the letter is known as the *coefficient*, and the number written after the variable that is raised is known as the *exponent*. An algebraic expression can be made up of one or more *terms*. Terms in algebra are separated by addition and subtraction. For example, the expression $3x^2$ has one term, the expression $4z - 5$ has two terms, and the expression $6y^3 + 4y - 2$ has three terms.

QUICK FACT

A simplified algebraic expression with one term is called a *monomial*, one with two terms it is called a *binomial*, and one with three terms it is called a *trinomial*. *Polynomial* is the general term for an algebraic expression containing one or more terms.

Practice

Identify each as a monomial, binomial, or trinomial.

1. $23x^4$

2. $g - 7$

3. y

4. $t^5 + 5t - 1$

5. $6t + 7u$

Be sure that a polynomial is in simplest form before classifying. For example, the polynomial $4x + 5x$ appears to be a binomial, but can be simplified to $9x$ by combining like terms. Therefore, this expression is a monomial. Like terms were explained in detail in Lesson 7.

Polynomial Operations

This section focuses on the operations of arithmetic (addition, subtraction, multiplication, and division) and how to apply them to polynomials.

Adding Polynomials

The key to adding polynomials is to be sure that only like terms are combined. Recall from Lesson 7 that like terms have exactly the same variable(s) with the same exponent(s). For example, the terms $7x^2$ and $-10x^2$ are like terms, but $5y^2$ and $5y^3$ are not like terms because the exponents are different. Remember that the coefficients are added or subtracted based on their signs and the variable part remains the same when combining like terms.

Example 1
Add $(4x + 3) + (12x - 8)$.

Combine the like terms of $(4x + 12x)$ and $(3 - 8)$ to get the simplified expression of $16x - 5$. The sum of the polynomials is $16x - 5$.

Example 2
Add $(x^2 - 3x - 10) + (-4x^2 - 6x + 12)$.

Combine the like terms of $(x^2 - 4x^2)$, $(-3x - 6x)$, and $(-10 + 12)$ to get the simplified expression of $-3x^2 - 9x + 2$. The sum of the polynomials is $-3x^2 - 9x + 2$.

Subtracting Polynomials

When subtracting polynomials, just as explained with integer arithmetic in Lesson 2, change the subtraction problem to add the opposite. In order to do this correctly, the signs of *all* the terms in parentheses that follow the subtraction sign must be changed to the opposite.

Example 1

$(3x^2 + 4x) - (2x^2 - 6x)$

Change the subtraction to addition, and change the signs of the terms in the second polynomial to their opposites. The problem becomes $(3x^2 + 4x) + (-2x^2 + 6x)$. Combine like terms to get the expression $3x^2 + -2x^2 + 4x + 6x$, which simplifies to $x^2 + 10x$.

Example 2

$(4m^2 - 3m + 10) - (-m^2 + 7m - 5)$

Change the subtraction to addition, and change the signs of the terms in the second polynomial to their opposites. The problem becomes $(4m^2 - 3m + 10) + (m^2 - 7m + 5)$. Combine like terms to get the expression $4m^2 + m^2 - 3m - 7m + 10 + 5$, which simplifies to $5m^2 - 10m + 15$.

Practice

Perform the indicated operation.

6. $(5y^3 + 12y) + (6y^3 - 3y)$

7. $(x^2 + 4x) + (-3x^2 - 12)$

8. $(-4x^2 + 5x - 8) + (-7x^2 - 9x - 1)$

9. $(8x^2 + 2x) - (7x^2 - 9x)$

10. $(-3y^2 - 7) - (-2y^2 - 4y - 2)$

11. $(-x^2 + 3x - 4) - (-x^2 - 9x - 10)$

Multiplying Polynomials

One of the keys to correctly multiplying polynomials is to know the rules for exponents from Lesson 6. To summarize, when multiplying like bases, add the exponents. The other key is to use the distributive property when the polynomials contain more than one term. There are four standard cases that can be studied.

Case 1: Multiplying a Monomial by a Monomial

As stated before, multiply coefficients and add the exponents on the like variables:

$$(4c^2)(-5c^6) = -20c^8$$

Case 2: Multiplying a Monomial by a Polynomial

Use the distributive property to multiply the monomial by all terms of the polynomial:

$$5t^3(3t^2 + t - 10) = 5t^3 \times 3t^2 + 5t^3 \times t - 5t^3 \times 10$$
$$= 15t^5 + 5t^4 - 50t^3$$

Case 3: Multiplying a Binomial by a Binomial

Use the distributive property to multiply both terms of the first binomial by each term in the second binomial. This can be remembered by using the acronym **FOIL** (first, outer, inner, last). For example, multiply $(x + 5)(x + 3)$ and then combine like terms.

Multiply the first terms of each binomial:
$x \times x = x^2$
Multiply the outer terms: $x \times 3 = 3x$
Multiply the inner terms: $5 \times x = 5x$
Multiply the last terms of each binomial:
$5 \times 3 = 15$

The product is $x^2 + 3x + 5x + 15$, which simplifies to $x^2 + 8x + 15$.

Case 4: Multiplying a Binomial by a Polynomial

Use the distributive property to multiply both terms of the first binomial by each term in the second polynomial:

$$(t^2 + 1)(t^3 + 2t - 3) = t^2 \times t^3 + t^2 \times 2t + t^2 \times -3 +$$
$$1 \times t^3 + 1 \times 2t + 1 \times -3$$
$$= t^5 + 2t^3 - 3t^2 + t^3 + 2t - 3$$

Combine like terms to simplify:

$$t^5 + 3t^3 - 3t^2 + 2t - 3$$

> ### TIP
>
> There are questions on placement exams that will ask you to "expand a polynomial." This is just asking you to multiply using the distributive property, as in the previous section.
>
> For example, to expand the binomial $(a + 3b)^2$, first write the binomial in parentheses as a factor the same number of times as the exponent. In this case, the binomial should be written as a factor twice:
>
> $(a + 3b)^2 = (a + 3b)(a + 3b)$
>
> Next, multiply using the distributive property. In this case, FOIL can be used:
>
> $(a + 3b)(a + 3b) = a^2 + 3ab + 3ab + 9b^2$
>
> Last, combine any like terms:
>
> $a^2 + 3ab + 3ab + 9b^2 = a^2 + 6ab + 9b^2$

Practice

Multiply the following.

12. $(8v^4)(10v)$

13. $(-2h^4)(-3h^9)$

14. $4x^3(x - 3)$

15. $-6y^2(-4y^2 + 5y)$

16. $(x + 3)(x + 2)$

17. $(y - 4)(2y + 1)$

18. $(x + 4)(x^2 - 5x + 2)$

19. Expand $(x + 1)^3$

Dividing Polynomials

To divide monomials, divide the coefficients and subtract the exponents of the like bases. The expression $\frac{4x^2}{2x}$ becomes $2x^1 = 2x$, the expression $\frac{16y^5}{4y^2}$ becomes $4y^3$, and the expression $\frac{10t^4s}{20ts}$ becomes $\frac{1}{2}t^3$.

To divide a polynomial by a monomial, divide each term in the numerator by the term in the denominator. For the problem $\frac{6x^2 + 4x}{2x}$, divide each term in the binomial by $2x$ and simplify each fraction separately. The problem becomes $\frac{6x^2}{2x} + \frac{4x}{2x} = 3x + 2$.

Practice

Simplify each of the following.

20. $\frac{12x^8}{4x^6}$

21. $\frac{-15g^{12}h}{-3g^{10}}$

22. $\frac{9x^3 + 3x}{3x}$

23. $\frac{20m^2 - 5m^2}{5m}$

Factoring Polynomials

The process of factoring polynomials is extremely useful in algebra and will be explained and refined here in this lesson. Factoring is a means to an end, so you will also see it used in other lessons, such as solving quadratic equations and simplifying rational expressions.

Factoring Using a Greatest Common Factor

When factoring using a greatest common factor, each of the terms has a number, a variable, or both in common. In the trinomial $2x^2 + 6x - 10$, each term contains a factor of 2. Place this factor of 2 to the left of parentheses, and in the parentheses write each term after a factor of 2 is divided out of each of the terms:

$$2x^2 + 6x - 10 = 2(x^2 + 3x - 5)$$

Factoring Using the Difference of Two Squares

The difference of two squares is a binomial that contains two perfect squares, separated by subtraction. Examples of this are $x^2 - 25$, $y^2 - 81$, and $4c^2 - 9e^4$. To factor the difference of two squares, the result is two binomials: the sum of the square roots and the difference. For example, the expression $x^2 - 25$ factors to $(x + 5)(x - 5)$, the expression $y^2 - 81$ factors to $(y + 9)$ $(y - 9)$, and the expression $4c^2 - 9e^4 = (2c + 3e^2)$ $(2c - 3e^2)$. To verify, you can FOIL the resulting binomials. The inner and outer terms always cancel each other out, leaving the difference of two squares.

TIP

The sum of two squares cannot be factored. For example, the binomial $x^2 + 4$ is not factorable even though x^2 and 4 are each perfect squares.

QUICK FACT

When a polynomial is in *standard form*, it is in order of decreasing exponents. A quadratic equation is an equation whose highest exponent is 2. The standard form of a quadratic equation is $ax^2 + bx + c$, where a, b, and c are real numbers. The polynomial $3x^2 + 5x - 7$ is in standard form and $a = 3$, $b = 5$, and $c = -7$.

Factoring Using the Sum/Product Rule

When factoring using the sum/product method, the polynomial will be a trinomial. First, be sure that the polynomial is in standard form, $ax^2 + bx + c$. Then, check to see if one pair of numbers has a sum of b and a product of c. For the trinomial $x^2 + 5x + 6$, the numbers 2 and 3 have a sum of 5 and a product of 6. The binomial factors are $(x + 2)$ and $(x + 3)$. Therefore $x^2 + 5x + 6 = (x + 2)(x + 3)$. In an additional example, the trinomial $y^2 - 3y - 18$ has values that have a sum of -3 and a product of -18. This pair of values is -6 and 3, so the factors become $(y - 6)(y + 3)$.

Practice

Factor each of the following polynomials.

24. $7x^2 + 14x$

25. $6a^3 - 3a + 6$

26. $x^2 - 49$

27. $25g^2 - 121$

28. $x^2 - 3x - 28$

29. $y^2 - 12y + 36$

Factoring $ax^2 + bx + c$ When $a > 1$

When given a trinomial to factor and the value of a is greater than 1, use a trial-and-error method with the factors of the leading term and the last term.

Example 1

For example, in order to factor the trinomial $3x^2 + 4x + 1$, possible factors of $3x^2$ are $3x$ and x, and 1 is the only factor of 1. Set up the parentheses with the factors for each: $(3x + 1)(x + 1)$. Now, check the result using FOIL. The check becomes $3x^2 + 3x + 1x + 1 = 3x^2 + 4x + 1$, which is the original trinomial.

Example 2

For the trinomial $2x^2 - 11x - 6$, set up the parentheses with the factors of $2x^2$ and -6. First try $(2x + 6)(x - 1)$ and check using FOIL. The result is $2x^2 - 2x + 6x - 6 = 2x^2 + 4x - 6$, which is not the same as the original trinomial. Switch the factors of -6 and try $(2x + 1)(x - 6)$ so that the 6 is multiplied by 2. Check using FOIL to get $2x^2 - 12x + 1x - 6 = 2x^2 - 11x - 6$. This is the same as the original trinomial, so the factors are $(2x + 1)(x - 6)$.

Practice

Factor each of the following polynomials.

30. $2x^2 - 5x - 3$

31. $3x^2 - 11x - 42$

32. $5x^2 + 39x - 8$

Factoring Completely

Sometimes, there can be more than one step in the factoring process. Polynomials should always be factored completely.

Examples

For example, in the binomial $3x^2 - 12$, the greatest common factor is 3. Factor out this common factor to get the product $3(x^2 - 4)$. The remaining binomial within the parentheses is a difference of two perfect squares, and can be factored further. The polynomial factored completely becomes $3(x - 2)(x + 2)$.

In the polynomial $5x^4 + 10x^3 - 240x^2$, factor out the greatest common factor of $5x^2$. The polynomial becomes $5x^2(x^2 + 2x - 48)$. The remaining trinomial in the parentheses can be

factored using the sum/product method. The values that have a sum of 2 and a product of -48 are 8 and -6. The final factored polynomial is $5x^2(x + 8)(x - 6)$.

Practice

Factor each of the following polynomials completely.

33. $4x^2 - 16$

34. $75x^5 - 27x^3$

35. $4x^4 - 8x^3 - 12x^2$

36. $3x^3 + 15x^2 + 12x$

Polynomial Graphs

The general nature of the graphs of polynomials can be determined by looking at the exponent and coefficient of the leading term of the polynomial. Recall that the degree of the term is the exponent, or the sum of the exponents if there is more than one variable.

Even-Degree Polynomials

If the degree of the leading term is an even value, the ends of the graph will go in the same direction. If the coefficient is positive, both ends go upward. If the coefficient is negative, both ends go downward. Two examples are shown here:

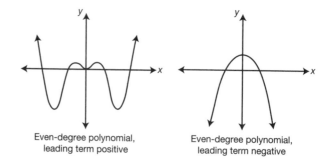

Even-degree polynomial, leading term positive

Even-degree polynomial, leading term negative

Odd-Degree Polynomials

If the degree of the leading term is an odd value, the ends of the graph will go in opposite directions. If the leading coefficient is positive, the graph enters on the left from the bottom and leaves the graph upward to the right. If the leading coefficient is negative, the graph enters on the left from the top, and leaves the graph downward to the right. These cases are shown here:

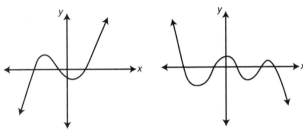

Odd-degree polynomial, leading term positive

Odd-degree polynomial, leading term negative

In describing the ends of the graph, the polynomial $y = x^4 + x^3 - 4$ has a leading coefficient with a degree of 4 (an even number) and a coefficient of 1 (a positive number). This graph has ends that go in the same direction and both ends point upward.

The polynomial $y = -2x^5 + x^4 + x^2 + 4$ has a leading coefficient with a degree of 5 (an odd number) and a coefficient of -2 (a negative number). This graph has ends that go in opposite directions and enters the graph on the left from the top, and exits the graph on the right downward.

Practice

Describe the end behavior of each of the following graphs.

37. $y = x^2$

38. $y = -x^2$

39. $y = x^5 + x^4 - x^3$

40. $y = -x^4 + x^2 - 10$

41. $y = -x^3 + x^2 - x - 5$

Practice Answers

1. monomial
2. binomial
3. monomial
4. trinomial
5. binomial
6. $11y^3 + 9y$
7. $-2x^2 + 4x - 12$
8. $-11x^2 - 4x - 9$
9. $x^2 + 11x$
10. $-y^2 + 4y - 5$
11. $12x + 6$
12. $80v^5$
13. $6h^{13}$
14. $4x^4 - 12x^3$
15. $24y^4 - 30y^3$
16. $x^2 + 5x + 6$
17. $2y^2 - 7y - 4$
18. $x^3 - x^2 - 18x + 8$
19. $x^3 + 3x^2 + 3x + 1$
20. $3x^2$
21. $5g^2h$
22. $3x^2 + 1$
23. $4m^3 - m$
24. $7x(x + 2)$
25. $3(2a^3 - a + 2)$
26. $(x + 7)(x - 7)$
27. $(5g + 11)(5g - 11)$
28. $(x - 7)(x + 4)$
29. $(y - 6)(y - 6)$
30. $(2x + 1)(x - 3)$
31. $(3x + 7)(x - 6)$
32. $(5x - 1)(x + 8)$
33. $4(x + 2)(x - 2)$
34. $3x^3(5x + 3)(5x - 3)$
35. $4x^2(x + 1)(x - 3)$
36. $3x(x + 4)(x + 1)$
37. The ends go in the same direction and point upward.
38. The ends go in the same direction and point downward.

39. The ends go in opposite directions and the graph enters from the bottom on the left and exits upward to the right.

40. The ends go in the same direction and point downward.

41. The ends go in opposite directions and the graph enters from the top on the left and exits downward to the right.

15 ▶ RATIONAL EXPRESSIONS AND EQUATIONS (INCLUDING QUADRATIC EQUATIONS)

For the COMPASS, ASSET, and ACCUPLACER

If a man empties his purse into his head, no man can take it away from him. An investment in knowledge always pays the best interest.

—BENJAMIN FRANKLIN

This chapter covers a lot of ground, and might take you longer than 20 minutes to complete. Split up your study into two parts if you need extra time to make your way through this lesson's examples.

LESSON SUMMARY

You are making an investment in knowledge by working through the lessons in this book to prepare for college placement exams. This lesson begins with the important procedure of solving quadratic equations. Then, it walks you through the steps to simplify rational expressions, and the cases where a rational expression is undefined. Operations with rational expressions are also covered and refined in this lesson. The process of solving equations with rational expressions concludes the lesson.

Solving Quadratic Equations

A quadratic equation is an equation that can be put in the form $ax^2 + bx + c = 0$, which is also called the standard form of the equation, as reviewed in Lesson 14. To solve a quadratic equation, first get the equation into standard form by setting the equation equal to zero. Then factor the polynomial, set each factor equal to zero, and solve for the variable.

Example 1
Solve the equation $x^2 = 9$ for all values of x.

First, put the equation in standard form and set the equation equal to zero. To do this, subtract 9 from both sides of the equation:

$x^2 - 9 = 0$

Next, factor the left side of the equation using the difference of two squares:

$(x - 3)(x + 3) = 0$

Then, set each factor equal to zero and solve for x. Recall that if the product of two factors equals zero, then at least one of the factors must equal zero.

$x - 3 = 0$ or $x + 3 = 0$

$x = 3$ or $x = -3$

Example 2
Solve the equation $x^2 + 2x = 24$ for all values of x.

First, subtract 24 from both sides of the equation:

$x^2 + 2x - 24 = 0$

Next, factor the left side of the equation using the sum/product method:

$(x - 4)(x + 6) = 0$

Then, set each factor equal to zero and solve for x:

$x - 4 = 0$ or $x + 6 = 0$

$x = 4$ or $x = -6$

If the equation is not factorable, use the quadratic formula.

DEFINITION

The *quadratic formula* is $x = \frac{-b \pm \sqrt{b^2 - 4ac}}{2a}$. To study this, try rewriting the formula each time you use it to solve an equation.

Take the following equation: $x^2 - x - 5 = 0$. This equation cannot be factored since there are no factors of -5 that have a sum of -1, so use the quadratic formula. First, be sure that the equation is in standard form, and identify a, b, and c. In this equation, $a = 1$, $b = -1$, and $c = -5$. Substitute into the quadratic formula and then simplify:

$$x = \frac{-b \pm \sqrt{b^2 - 4ac}}{2a} = \frac{-(-1) \pm \sqrt{(-1)^2 - 4(1)(-5)}}{2(1)}$$

$$= \frac{1 \pm \sqrt{1 + 20}}{2} = \frac{1 \pm \sqrt{21}}{2}$$

The solution can be simplified to $\frac{1}{2} + \frac{\sqrt{21}}{2}$ or $\frac{1}{2} - \frac{\sqrt{21}}{2}$, which approximates to 2.79 or -1.79.

DID YOU KNOW?

All quadratic equations can be solved using the quadratic formula, whether they are factorable or not.

Practice
Solve each of the following quadratic equations for all values of x.

1. $x^2 + 3x + 2 = 0$

2. $x^2 = 25$

3. $x^2 - 4x = 21$

4. $x^2 + 6x + 6 = 0$

5. $x^2 + 3x - 10 = 0$

Simplifying Rational Expressions

DEFINITION

A *rational expression* is a fraction where the numerator and denominator are polynomials.

The key to simplifying rational expressions is to be sure that the numerator and denominator are factored completely first. Recall the various techniques for factoring polynomials from the previous lesson to assist with this.

Take, for example, the expression $\frac{2x+4}{4x+8}$. First, factor both numerator and denominator using the greatest common factor: $\frac{2(x+2)}{4(x+2)}$.

Cancel the common factors of 2 and $(x + 2)$ to get the simplified expression:

$$\frac{\overset{1}{\cancel{2}}\ \overset{1}{\cancel{(x+2)}}}{\underset{2}{\cancel{4}}\ \underset{1}{\cancel{(x+2)}}} = \frac{1}{2}$$

QUICK FACT

When canceling common factors, the factor is replaced with a 1, not a zero.

For another example, take the expression $\frac{x^2-9}{x^2+5x+6}$. First, factor the numerator using the difference of two squares and then factor the denominator using the sum/product rule:

$$\frac{(x+3)(x-3)}{(x+3)(x+2)}$$

Cancel the common factor of $x + 3$ from both the numerator and denominator:

$$\frac{\overset{1}{\cancel{(x+3)}}(x-3)}{\underset{1}{\cancel{(x+3)}}(x+2)} = \frac{x-3}{x+2}$$

TIP

When simplifying rational expressions, cancel out *only* common factors. In the expression $\frac{x-3}{x+2}$, the x terms cannot be canceled out because they are not factors, so the expression is simplified. Remember that factors are multiplied, not added or subtracted.

To simplify an expression such as $\frac{x-3}{3-x}$, first put the denominator in standard form and factor out a -1:

$$\frac{x-3}{-x+3} = \frac{x-3}{-1(x-3)}$$

Cancel the common factors and replace them with a 1:

$$\frac{\overset{1}{\cancel{(x-3)}}}{-1\underset{1}{\cancel{(x-3)}}} = \frac{1}{-1} = -1$$

TIP

When an expression such as $\frac{x-5}{5-x}$ is simplified, the result is -1.

Another important example is the expression $\frac{x-6}{36-x^2}$. In order to factor this expression, first put the denominator in standard form, and then factor out a -1:

$$\frac{x-6}{-x^2+36} = \frac{x-6}{-1(x^2-36)}$$

Next, factor the binomial in parentheses using the difference of two squares:

$$\frac{x-6}{-1(x-6)(x+6)}$$

Cancel the common factor of $x-6$ to get the simplified expression:

$$\frac{\overset{1}{\cancel{(x-6)}}}{-1\underset{1}{\cancel{(x-6)}}(x+6)} = \frac{1}{-1(x+6)} = -\frac{1}{x+6}$$

Practice

Simplify each of the following rational expressions.

6. $\frac{3x+6}{5x+10}$

7. $\frac{2x^2+6x}{4x+12}$

8. $\frac{x^2-49}{x^2+2x-63}$

9. $\frac{x^2+4x+4}{x^2+7x+10}$

10. $\frac{x-8}{8-x}$

11. $\frac{x-4}{16-x^2}$

When a Rational Expression Is Undefined

Because dividing by zero is undefined, a rational expression, just like a regular fraction, cannot have a denominator of zero. Often, we must determine for what values of the variables the expression will be undefined.

QUICK FACT

The denominator of the rational expression determines when the expression is undefined.

Example

For what value of x is the expression $\frac{1}{x-2}$ undefined?

To determine the value, first set the denominator equal to zero. Then, solve for x:

$x - 2 = 0$

Add 2 to each side of the equation:

$x - 2 + 2 = 0 + 2: x = 2$

This expression is undefined when the value of x is 2.

In the expression $\frac{4x^2}{x^2+6x-27}$, set the denominator equal to zero and solve for all values of x:

$x^2 + 6x - 27 = 0$

Factor the left side of the equation to get $(x+9)(x-3) = 0$. Set each factor equal to zero: $x + 9 = 0$ or $x - 3 = 0$, so $x = -9$ or 3.

Practice

Determine for which value(s) of x the following expressions are undefined.

12. $\frac{1}{x}$

13. $\frac{3}{x-9}$

14. $\frac{6x}{x+6}$

15. $\frac{x+7}{x^2+x-20}$

Operations with Rational Expressions

Each of the four operations with rational expressions can be related to the operations with rational numbers, as explained in Lesson 3. Refer to that lesson for a refresher on these topics before beginning this section, if necessary.

Addition of Rational Expressions

To add rational expressions, first get a common denominator. Then, add the numerators as usual.

To add the expressions $\frac{1}{3x} + \frac{2}{9x^2}$, first get the least common denominator of $9x^2$. Multiply the numerator and denominator of the first expression by $3x$ to convert to the common denominator:

$$\frac{1 \times 3x}{3x \times 3x} + \frac{2}{9x^2}$$

The problem becomes $\frac{3x}{9x^2} + \frac{2}{9x^2}$.

Combine the numerators and keep the common denominator:

$$\frac{3x + 2}{9x^2}$$

Subtraction of Rational Expressions

To subtract rational expressions, first get a common denominator. Then, subtract the numerators as usual. Take the following example.

Find the difference:

$$\frac{4}{x - 3} - \frac{x}{x^2 - 9}$$

First, multiply each term by a factor to convert to a common denominator. Recall that the denominator of $x^2 - 9$ factors to $(x + 3)(x - 3)$. The expression becomes $\frac{4(x + 3)}{(x - 3)(x + 3)} - \frac{x}{(x - 3)(x + 3)}$. Subtract the numerators and keep the common denominator:

$$\frac{4x + 12 - x}{(x - 3)(x + 3)}$$

Simplify to get $\frac{3x + 12}{(x - 3)(x + 3)} = \frac{3x + 12}{x^2 - 9}$.

Multiplication of Rational Expressions

To multiply rational expressions, first check to be sure that all numerators and denominators are factored completely. Cancel out any common factors. Then, multiply the numerators together and multiply the denominators together as usual; check to be sure that the final expression is simplified.

Multiply and simplify:

$$\frac{x^2 - 25}{x + 8} \times \frac{x^2 + 8x}{x^2 - 5x}$$

First, factor each numerator and denominator, if possible. The expression then becomes $\frac{(x - 5)(x + 5)}{x + 8} \times \frac{x(x + 8)}{x(x - 5)}$. Cancel any common factors, with one in the numerator and one in the denominator, regardless of which expression they originally came from:

$$\frac{\cancel{(x - 5)}(x + 5)}{\cancel{x + 8}} \times \frac{\cancel{x}\cancel{(x + 8)}}{\cancel{x}\cancel{(x - 5)}} = x + 5$$

Division of Rational Expressions

To divide rational expressions, first change the problem to multiplication and take the reciprocal of the divisor. Then follow the steps described earlier for multiplication.

Perform the division and simplify:

$$\frac{4x - 20}{2x} \div \frac{x^2 - 25}{10x^2}$$

First, change the division to multiplication and take the reciprocal of the fraction following the division. The expression becomes $\frac{4x - 20}{2x} \times \frac{10x^2}{x^2 - 25}$.

Factor each numerator and denominator:

$$\frac{4(x - 5)}{2x} \times \frac{10x^2}{(x - 5)(x + 5)}$$

Cancel any common factors, one from the numerator and one from the denominator:

$$\frac{\overset{2}{\cancel{4}}\cancel{(x - 5)}}{\underset{1}{\cancel{2x}}} \times \frac{10x\overset{1}{\cancel{x}}}{\cancel{(x - 5)}(x + 5)}$$

Multiply the remaining factors from the numerator and denominator. The simplified quotient is $\frac{20x}{x + 5}$.

QUICK FACTS

When adding and subtracting rational expressions, a common denominator is necessary.

When multiplying and dividing rational expressions, a common denominator is *not* necessary.

Practice

For each of the following, perform the indicated operation.

16. $\frac{5}{x} + \frac{7}{2x}$

17. $\frac{3}{x} + \frac{x+1}{x+2}$

18. $\frac{2x}{5} - \frac{x-1}{5x}$

19. $\frac{x+2}{x-2} - \frac{3}{2x-4}$

20. $\frac{x+2}{3x-9} \times \frac{x^2-9}{10x+20}$

21. $\frac{4x+28}{16x^2} \times \frac{32x^3}{x^2+4x-21}$

22. $\frac{x}{x+5} \div \frac{3x}{x+5}$

23. $\frac{81-x^2}{4x} \div \frac{x^2-9x}{4x-32}$

Solving Rational Equations

When solving rational equations, there are a few different ways they can be solved. In the first case, you can find a common denominator for all the terms in the equation, and then follow the rules for rational numbers and solving equations.

Example 1

Solve for x: $\frac{x}{5} + \frac{3x}{10} = -4$.

First, convert each term in the equation to the least common denominator of 10. The equation becomes $\frac{2x}{10} + \frac{3x}{10} = \frac{-40}{10}$.

Combine the like terms on the left side of the equation: $\frac{5x}{10} = \frac{-40}{10}$.

Because the denominators are equal and multiplying each side by 10 cancels the denominators, set the numerators equal to each other:

$5x = -40$, so $x = -8$

In the second method, the expressions in the equation can be cross multiplied and the equation solved for the variable.

Example 2

Solve for x: $\frac{x+4}{3} = \frac{4}{x}$.

Find the cross products and set them equal to each other:

$x^2 + 4x = 12$

Subtract 12 from each side to get the equation in standard form:

$x^2 + 4x - 12 = 0$

Factor the left side of the equation into two binomials:

$(x + 6)(x - 2) = 0$

Set each factor equal to zero and solve for x:

$x + 6 = 0$ or $x - 2 = 0$

$x = -6$ or $x = 2$

TIP

A quick way to eliminate the fractions in a rational equation is to multiply all terms of the equation by the least common denominator.

Another way to efficiently solve rational equations is mentioned in the preceding tip: multiply each of the terms by the least common denominator in the first step. The remaining equation should not have any fractions after that, and should be easier to solve.

Example

Solve for x: $\frac{11}{4x} - \frac{2}{3x} = \frac{1}{6}$.

First, multiply each term of the equation by the least common denominator of $12x$. The equation becomes $\cancel{12x}^{3} \times \frac{11}{\cancel{4x}} - \cancel{12x}^{4} \times \frac{2}{\cancel{3x}} = \cancel{12x}^{2x} \times \frac{1}{\cancel{6}}$, which simplifies to $33 - 8 = 2x$.

Further simplify to get $25 = 2x$. Divide each side of the equation by 2 to get $12.5 = x$.

Practice

Solve each of the following rational equations for the indicated variable.

24. $\frac{1}{2} + \frac{5}{x} = \frac{8}{x}$

25. $\frac{2x+1}{5} - \frac{2+7x}{15} = \frac{4}{6}$

26. $\frac{x-1}{6} = \frac{x+2}{9}$

27. $\frac{x}{4} - \frac{1}{2} + \frac{x}{6} = 7$

28. $x + \frac{x}{x-1} = \frac{4x-3}{x-1}$

Solving Rational Inequalities

Solving rational inequalities is the same basic procedure as solving rational equations. Recall the information on inequalities from Lesson 9 for assistance with inequalities.

Example

Solve for x: $\frac{1}{3x} + \frac{1}{x} \geq 2$.

First, multiply each term in the inequality by the least common denominator of $3x$ and cancel any common factors:

$$\cancel{3x} \frac{1}{\cancel{3x}} + 3\cancel{x} \frac{1}{\cancel{x}} \geq 3x \times 2$$

The inequality becomes $1 + 3 \geq 6x$.

Simplify to get $4 \geq 6x$.

Divide each side by 6 to get $\frac{4}{6} \geq \frac{6x}{6}$.

Simplify to get a solution of $\frac{2}{3} \geq x$.

Practice

Solve each of the following inequalities.

29. $\frac{40}{x} \leq 10$

30. $\frac{2}{x} + 3 > \frac{-4}{x}$

31. $\frac{3}{x+2} < \frac{4}{x-1}$

Practice Answers

1. -1 or -2
2. -5 or 5
3. -3 or 7
4. Exact: $-3 + \sqrt{2}$ or $-3 - \sqrt{2}$; approximate: -4.73 or -1.27
5. -5 or 2
6. $\frac{3}{5}$
7. $\frac{x}{2}$
8. $\frac{x+7}{x+9}$
9. $\frac{x+2}{x+5}$
10. -1
11. $-\frac{1}{x+4}$
12. 0
13. 9
14. -6
15. -5 or 4

16. $\frac{17}{2x}$
17. $\frac{x^2 + 4x + 6}{x(x+2)} = \frac{x^2 + 4x + 6}{x^2 + 2x}$
18. $\frac{2x^2 - x + 1}{5x}$
19. $\frac{2x + 1}{2(x+2)} = \frac{2x + 1}{2x + 4}$
20. $\frac{x + 3}{30}$
21. $\frac{8x}{x - 3}$
22. $\frac{1}{3}$
23. $-\frac{(x+9)(x-8)}{x^2} = \frac{x^2 + x - 72}{x^2}$
24. $x = 6$
25. $x = -9$
26. $x = 7$
27. $x = 18$
28. $x = 3 \ (x \neq 1)$
29. $x \geq 4$
30. $x > -2$
31. $x > -11$

16▶

FUNCTIONS
For the COMPASS, ASSET, and ACCUPLACER

A person who never made a mistake never tried anything new.
—ALBERT EINSTEIN

LESSON SUMMARY
This lesson reviews the important concept of function. The general definition of function as a verb is to "work or operate in a particular way." A function is a special equation in two variables that behaves in a particular way. Along with reviewing functions, you will learn about the concepts of domain and range, function notation, and function operations such as composition and inverse.

Function

A function is a relationship between a set of numbers, called *inputs*, and another set of numbers, called *outputs*. Often, functions are shown as a set of ordered pairs, as reviewed in Lesson 12, and can be graphed. With a function, each and every input has exactly one output. Two different inputs can yield the same output, but each input has *only* one output.

Usually, the input values are denoted as x and the output values are denoted as $f(x)$, which is read as "f of x" or "a function of x." The following figure shows examples of relationships that are functions and a relationship that is not a function.

x	f(x)
–3	10
–2	12
–1	14
0	16
1	18
2	20
3	22

Function
{(–3,10), (–2,12), (–1,14), (0,16)
(1,18), (2,20), (3,22)}

x	f(x)
–3	8
–2	6
–2	4
–1	3
0	2
1	0
2	–2

Not a Function
{(–3,8), (–2,6), (–2,4), (–1,3)
(0,2), (1,0), (–2, 2)}

DEFINITIONS

The *domain* of a function is the set of possible input values. It may or may not be an infinite set.
The *range* of a function is the set of output values. It also may or may not be an infinite set.

The table on the left shows a function. Each and every *x* value has a single *f(x)* value. The table on the right is not a function. The *x* value of –2 has two separate *f(x)* values of 6 and 4. Many equations in two variables are functions, if each *x*-coordinate has one *y*-coordinate. You can determine if a graphed equation is a function by performing a vertical line test.

In the relationship {(–3,1), (–2,5), (–1,9), (0,13)}, the domain is {–3, –2, –1, 0} and the range is {1, 5, 9, 13}. In the relationship defined by $f(x) = 7x - 3$, both the domain and the range are all the real numbers.

Practice

Determine whether the relationships shown here are functions. Then state the domain and range for each.

1. {(–3,20), (–2,17), (–1,14), (0,11), (1,8), (2,5)}

2. {(6,5), (7,4), (8,3), (6,2), (9,1), (10,0)}

3. {(2,4), (3,4), (4,5), (5,2), (6,0), (7,9)}

QUICK FACT

To determine whether a graphed equation with real values of *x* is a function, perform a vertical line test. Place a vertical line anywhere across the *x*-axis, and verify that it produces one and only one point that intersects with the equation.

Look at the following graphs. On the left, the relationship is a function. The vertical line test holds. The relationship on the right is not a function. Note that the vertical line intersects the graph of the equation in two places.

4.

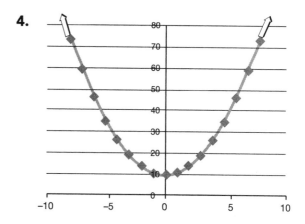

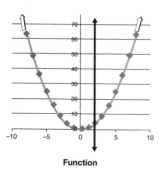

Function

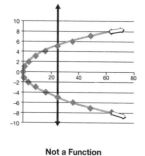

Not a Function

5.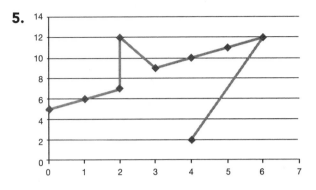

Function Notation and Evaluating

You reviewed linear equations in Lesson 12. Most of these linear equations are functions, for example $y = 3x + 2$. When using function notation, instead of referring to the second variable as y, it is referred to as $f(x)$, which indicates that it is in fact a function.

To find a specific value of the function for a specific x, you substitute the value of x into the function rule and evaluate the expression by following the order of operations. This is evaluating the function. So, for example, to find $f(4)$ when $f(x) = 3x + 2$, substitute 4 in to get $3(4) + 2 = 12 + 2 = 14$. So it is said that $f(4) = 14$. The function does not have to be named f. Any letter of the alphabet is acceptable.

Practice
Evaluate the function $f(x) = 4x - 10$:

6. $f(5)$

7. $f\left(\frac{2}{3}\right)$

8. $f(-3)$

Evaluate the function $h(x) = x^2 + 6x + 9$:

9. $h(-3)$

10. $h\left(\frac{1}{2}\right)$

Composition of Functions

A composition of functions refers to when you apply a second function to the output of the first function. Similar to the way you evaluate parentheses in the order of operations, you do the inner function first, and then the outer function.

Example 1
For example, let one function be defined as $f(x) = 2x + 6$ and another function be defined as $g(x) = x^2 - 3$. So to evaluate the composition $g[f(5)]$, first evaluate $f(5)$:
$$f(5) = 2(5) + 6 = 16$$
Now take this result, 16, and use it as the input in the other function:
$$g(16) = 16^2 - 3 = 256 - 3 = 253$$

Another way to express a composition of functions is with the dot operator: ∘. So the preceding composition can be written as $g \circ f(5)$. In this notation, the rightmost function is evaluated first.

Example 2
If $f(x) = -\frac{1}{4}x + 4$ and $g(x) = 6x - 2$, then to evaluate $f \circ g(4)$, first calculate $g(x)$:

$$g(x) = 6(4) - 2 = 24 - 2 = 22$$

Now use 22 as the input in the f function:

$$f(x) = -\frac{22}{4} + 4 = -5.5 + 4 = -1.5$$

Practice
Given $f(x) = 2x^2 + 6x - 8$, $g(x) = -4x + 2$, and $h(x) = 3x - 10$, evaluate:

11. $g[f(2)]$

12. $f \circ g(2)$

13. $g[h(10)]$

14. $h \circ g(10)$

15. $f[h(5)]$

16. $h \circ f(5)$

Inverse of Functions

A function takes an input, x, and produces an output, $f(x)$, or y. The *inverse* of a function treats the output as the input, and vice versa. The notation for the inverse of the function $f(x)$ is $f'(x)$ or alternatively, $f^{-1}(x)$.

The inverse is the rule that would take the output and produce the input. If the function is a set of ordered pairs, such as $f(x) = \{(0,1), (2,3), (4,5), (6,7)\}$, then the inverse function, $f'(x)$, is $\{(1,0), (3,2), (5,4), (7,6)\}$.

Example
When given a function such as $f(x) = -3x + 8$, to determine the inverse, $f^{-1}(x)$, first rewrite it as $y = -3x + 8$, exchange the x and the y, and solve for y:

$y = -3x + 8$ Write the original function.

$x = -3y + 8$ Exchange the x and y and solve for y.

$\underline{-8 = \qquad -8}$ Subtract 8 from each side.

$x - 8 = -3y$ Simplify.

$\frac{x-8}{-3} = \frac{-3y}{-3}$ Divide both sides by -3.

$-\frac{x-8}{-3} = y$ This is $f^{-1}(x)$, the inverse of the function.

Sometimes the inverse of a function is also a function, in which case it is called a one-to-one function. Other times, the inverse is not a function, just a relation.

Practice
Find the inverse of these functions. Then, state whether this inverse is a function:

17. $f(x) = \{(5,-2), (6,-1), (7,0), (8,-1), (9,-2)\}$

18. $f(x) = \{(5,12), (6,14), (7,16), (8,18), (9,20)\}$

19. $f(x) = 10x - 15$

20. $f(x) = \frac{2}{3}x + 18$

21. $f(x) = 36x^2$

Practice Answers

1. Yes, it is a function. The domain is {−3, −2, −1, 0, 1, 2}. The range is {20, 17, 14, 11, 8, 5}.

2. No, it is not a function. The domain is {6, 7, 8, 9, 10}. The range is {5, 4, 3, 2, 1, 0}.

3. Yes, it is a function. The domain is {2, 3, 4, 5, 6, 7}. The range is {0, 2, 4, 5, 9}.

4. Yes, it passes the vertical line test. The domain is all of the real numbers. The range is $\{y \mid y \mid \geq 10\}$.

5. No, it does not pass the vertical line test. The domain is $\{x \mid 0 \leq x \leq 6$. The range is $\{y \mid 2 \leq y \leq 12\}$.

6. 10

7. $-\frac{22}{3}$ or $-7\frac{1}{3}$

8. −22

9. 0

10. $\frac{49}{4}$ or $12\frac{1}{4}$

11. −46

12. 28

13. −78

14. −124

15. 72

16. 206

17. $f'(x) = \{(-2,5), (-1,6), (0,7), (-1,8), (-2,9)\}$. The inverse is not a function.

18. $f'(x) = \{(12,5), (14,6), (16,7), (18,8), (20,9)\}$. The inverse is a function.

19. $f'(x) = \frac{x+15}{10}$. The inverse is a function.

20. $f'(x) = \frac{3}{2}(x-18) = \frac{3x}{2} - 27$. The inverse is a function.

21. $f'(x) = \pm\sqrt{\frac{x}{36}} = \pm\frac{\sqrt{x}}{6}$. The inverse is not a function.

17 ▶ EXPONENTIAL FUNCTIONS AND LOGARITHMS
For the COMPASS, ASSET, and ACCUPLACER

Whatever your difficulties in mathematics, I can assure you mine are far greater.

—ALBERT EINSTEIN

LESSON SUMMARY
You reviewed functions in the previous lesson. This lesson reviews two special functions—the exponential function and its inverse, the logarithm function. You learned about the Laws of Exponents in Lesson 6; these are reviewed again, along with the corresponding Laws of Logarithms. You will also learn how to create and recognize the graph of an exponential function. Finally, you will recall how to solve exponential equations.

Exponential Functions

The last lesson covered linear functions. There are other types of functions, such as polynomial functions, like $y = x^2 + 4x + 4$, or $y = x^3$. Exponential functions are those that have the variable as the exponent. For example, $f(x) = 10^x$ is an exponential function. The domain of the function is all real numbers. The range of the function is $y > 0$.

For every exponential function $y = a \circ b^x$, the base, b, is always positive. The range of these function types is also always positive. Also, recall that a negative exponent does not make a number negative; when b is a number greater than or equal to 1, b^{-x} will be a fraction, but never negative.

Practice

Identify whether the relation is an exponential function. If not, say why.

1. $f(x) = 5^x$

2. $f(x) = x^5 + 12$

3. $f(x) = (-5)^x$

Evaluate:

4. $f(0)$ when $f(x) = 6^x$

5. $f(5)$ when $f(x) = 2^x$

6. $f(4)$ when $f(x) = 3^x + 10$

Laws of Exponents

The Laws of Exponents were reviewed in Lesson 6. In those examples, the base was a variable and the exponents were constants. These same laws apply when the base is a constant and the exponents are variables.

QUICK FACTS

For $b > 0$, and for any x and y:

$b^x b^y = b^{x+y}$

$\dfrac{b^x}{b^y} = b^{x-y}$

$b^0 = 1$

$b^{-x} = \dfrac{1}{b^x} = \left(\dfrac{1}{b}\right)^x$

$(b^x)^y = b^{xy}$

Example

To evaluate $(25^{\frac{1}{3}})^3$, you do not need to find the cube root of 25, but you can use the Laws of Exponents and multiply the powers:

$$(25^{\frac{1}{3}})^3 = 25^{\frac{1}{3} \times 3} = 25$$

Practice

Simplify:

7. $\dfrac{3^4 + 3^5}{3^6}$

8. $(8^{\frac{1}{2}})^4$

9. $(27^{\frac{3}{4}})^0$

Exponential Equations

Equations with Exponents That Are Either Fractions or Negative

Recall that fractional and negative exponents were reviewed in Lesson 6. If you encounter an equation with either a fractional or negative exponent, first isolate the base variable, and then perform an exponent to each side of the equation to make the fraction or negative exponent equal to 1 (or at the very least positive).

Example

To solve $4x^{\frac{2}{3}} = 64$:

$\dfrac{4x^{\frac{2}{3}}}{4} = \dfrac{64}{4}$ Divide both sides to first isolate x.

$x^{\frac{2}{3}} = 16$ Simplify.

$(x^{\frac{2}{3}})^{\frac{3}{2}} = 16^{\frac{3}{2}}$ Take both sides of the equation to the $\frac{3}{2}$ power.

$x^1 = (\sqrt{16})^3$ Rewrite the fraction exponent.

$x = 64$ Simplify.

Use a similar technique if the exponent is negative:

$5x^{-\frac{1}{2}} = 15$	Write the original equation.
$\frac{5x^{-\frac{1}{2}}}{5} = \frac{15}{5}$	Divide both sides by 5 to isolate x.
$x^{-\frac{1}{2}} = 3$	Simplify.
$(x^{-\frac{1}{2}})^{-2} = (3)^{-2}$	Take both sides of the equation to the –2 power.
$x^1 = 3^{-2}$	
$x = \frac{1}{9}$	Simplify.

Practice

Solve, given that the variables are positive numbers.

10. $x^{\frac{2}{3}} - 1 = 15$

11. $x^{-\frac{1}{5}} = 2$

12. $3x^{\frac{4}{3}} = 48$

Equations with a Variable Exponent and Like Bases

To solve an equation with like bases, you can simply realize that the exponents are then also equal. For example, to solve $8^{3x-5} = 8^{10}$, note that the bases are the same (8). Since the two sides of the equation are equal, the exponents must have the same value. You can then make the equation $3x - 5 = 10$ and solve for x. Add 5 to both sides to get $3x - 5 + 5 = 10 + 5$, or $3x = 15$. Divide both sides by 3, and $x = 5$.

Sometimes the bases are not the same, but one base may be a power of the other. In that case, change them to have the same base.

Example

$2^{6x+3} = 8^7$	Write the original equation.
$2^{6x+3} = (2^3)^7$	Recognize and rewrite 8 as 2^3.
$2^{6x+3} = 2^{21}$	Apply the appropriate law of exponents.
$6x + 3 = 21$	Bases are now the same, so make an equation of the exponents.
$\underline{-3 = -3}$	Subtract 3 from both sides.
$6x = 18$	
$\frac{6x}{6} = \frac{18}{6}$	Divide both sides by 6.
$x = 3$	Simplify.

Practice

Solve:

13. $5^{4x-5} = 5^{x+10}$

14. $3^{5x+1} = 9^x$

15. $2^{2x+10} = 4^{10}$

The Graph of an Exponential Function

The graph of the exponential function $f(x) = 2x$ is shown here.

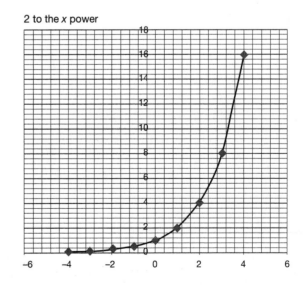

2 to the x power

The nature of this exponential function graph and all exponential functions (with a base > 0) is that it is always positive. It starts out very small, hugging the x-axis when $x < 0$. Recall that this is because a negative exponent will still yield a positive fraction when the base is positive. Note that the graph passes through the point (0,1), and remember that will be the case because any number to the zero power is 1. Once the x-values are positive, the graph rises sharply and quickly. You can see the behavior of the function in the following figure, which zooms in on small values of x.

17.

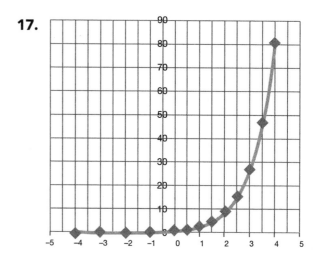

2 to the x power zoomed

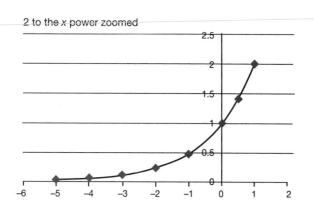

18. 2 to the x power

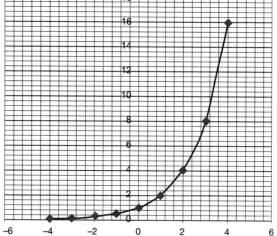

Practice

Identify whether these graphs are exponential functions.

19.

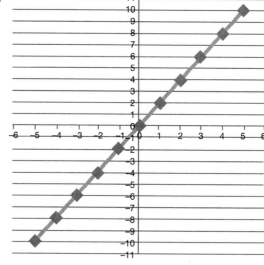

16.

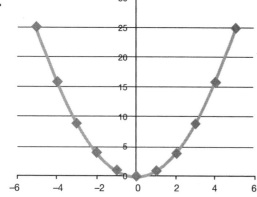

Logarithms

Logarithms are exponents. A logarithm is the inverse of an exponential function. For example, when $10^x = y$, it is said that the logarithm, or log, of the number y is the exponent x. You can take any exponential equation and show it in logarithmic form.

DEFINITION

The relationship between the exponential function and the logarithmic function using a base of 10 is given by:

$10^1 = 10$ and $\log 10 = 1$

$10^2 = 100$ and $\log 100 = 2$

$10^3 = 1{,}000$ and $\log 1{,}000 = 3$

$10^{-3} = 0.001$ and $\log 0.001 = -3$

Because we operate in the decimal system, the base of 10 is called the *common logarithm*. Scientific calculators have a "log" key that is in the base of 10. When you refer to bases other than 10, you specify the base as a subscript to the word *log*. For example, because $2^5 = 32$, then $\log_2 32 = 5$.

Practice

Evaluate:

20. $\log_3 9$

21. $\log_5 125$

Find the value of x.

22. $\log_4 x = 3$

23. $\log_4 8 = 3$

Laws of Logarithms

Because logarithms are the inverse of exponentials, the Laws of Logarithms are closely related to the Laws of Exponents.

QUICK FACTS

Because $b^x b^y = b^{x+y}$, then $\log_b(cd) = \log_b c + \log_b d$

Because $\frac{b^x}{b^y} = b^{x-y}$, then $\log_b(\frac{c}{d}) = \log_b c - \log_b d$

Because $b^0 = 1$, then $\log_b 1 = 0$

Because $b^{-x} = \frac{1}{b^x} = (\frac{1}{b})^x$, then $\log_b(\frac{1}{c}) = -\log_b c$

Because $(b^x)^y = b^{xy}$, then $\log(c^x) = x\log c$

Using the Laws of Logarithms helps to solve problems that otherwise could not be solved easily.

Example

To find the value of $\log_8 32 + \log_8 2$, you do not immediately know what power of 8 would yield 32, or what power of 8 would yield 2. These would be fractional. But if you apply the Laws of Logarithms you can easily solve this:

$\log_8 32 + \log_8 2$	Write the original expression.
$\log_8 32 \times 2$	Rewrite using the Laws of Logarithms.
$\log_8 64$	Simplify.
$8^x = 64$	Rewrite as an exponential.
$x = 2$	The exponent is 2 because $8 \times 8 = 64$.

To solve for x in the equation, again use several of the Laws of Logarithms:

$2\log_5 10 + \log_5 x = \log_5 50$	Write the original equation.
$\log_5 10^2 + \log_5 x = \log_5 50$	Use the Power Rule of Logarithms.
$\log_5 100 + \log_5 x = \log_5 50$	Simplify.
$\log_5 100x = \log_5 50$	Use the Multiplication Rule of Logarithms.
$100x = 50$	Because both sides refer to a log in the base of 5.
$\frac{100x}{100} = \frac{50}{100}$	Divide both sides by 100 to isolate x.
$x = \frac{1}{2}$	

Practice

Evaluate:

24. $\frac{\log_2 1}{8} + \log_2 256$

25. $\log_4 160 - \frac{1}{2}\log_4 100$

Solve for n.

26. $3\log_3 4 = \log_3 n$

27. $\log_2 n - \frac{\log_2 1}{6} = \log_2 54$

Using Logarithms to Solve an Exponential Equation

Earlier in this lesson, we reviewed how to solve an exponential equation with like bases, or like powers of the same base. To solve an exponential equation where the bases are not similar powers, you can use common logarithms.

Example

To solve $6^x = 14$ to the nearest hundredth, note that 6 and 14 do not have similar powers. So take the common logarithm of each side and use the Laws of Logarithms:

$6^x = 14$	Write the original equation.
$\log 6^x = \log 14$	Take the common logarithm of each side.
$x \log 6 = \log 14$	Use the Power Rule of Logarithms.
$x\frac{\log 6}{\log 6} = \frac{\log 14}{\log 6}$	Divide both sides of the equation by $\log 6$ to isolate x.
$x \approx \frac{1.15}{0.78}$	Use a scientific or graphing calculator to find the common logarithms.
$x \approx 1.47$	Divide and round the answer.

You can check this equation by using a calculator to verify that $6^{1.47} \approx 14$.

If you are given an equation in logarithmic form, change the equation to an exponential form. To solve $\log_5 25 = x$, change the equation to $5^x - 25$. You can most likely solve this equation by inspection; that is, you can identify that $x = 2$, based on your knowledge of the multiplication tables.

If you change the equation to exponential form and cannot determine the answer by inspection, you will need to use logarithms.

Example

Solve $\log_3 16 = x$ to the nearest tenth:

$\log_3 16 = x$	Write the original equation.
$3^x = 16$	Rewrite as an exponential equation.
$\log 3^x = \log 16$	Take the common logarithm of each term on both sides.
$x \log 3 = \log 16$	Use the Power Rules of Logarithms.
$\frac{x \log 3}{\log 3} = \frac{\log 16}{\log 3}$	Divide both sides by $\log 3$.
$x \approx \frac{1.20}{4.8}$	Use the calculator to find the common logarithms.
$x \approx 2.5$	Divide and round the answer.

QUICK FACT

You can take the common logarithm of each term of both sides of an equation to make an equivalent equation.

Practice

Solve for x. Round to the nearest tenth if necessary.

28. $8^x = 24$

29. $12^x = 200$

30. $6^x = 3$

31. $\log_7 1 = x$

32. $\log_3 18 = x$

Practice Answers

1. Yes

2. No; exponential functions have the variable as the exponent.

3. No; the base in an exponential function is always positive.

4. 1

5. 32

6. 91

7. $3^3 = 27$

8. 64

9. 1

10. 64

11. $\frac{1}{32}$

12. 8

13. $x = 5$

14. $x = -\frac{1}{3}$

15. $x = 5$

16. No; this is the quadratic function $y = x^2$.

17. Yes; $y = 3^x$

18. Yes; $y = 2^x$

19. No; this is the linear function $y = 2x$.

20. 2

21. 3

22. 64

23. 2

24. $\log_2 32 = 5$

25. $\log_4 16 = 2$

26. $n = 4^3 = 64$

27. $n = 9$

28. $x \approx 1.5$

29. $x \approx 2.1$

30. $x \approx 0.6$

31. $x = 0$

32. $x \approx 2.6$

18 ▶ TRIGONOMETRY For the COMPASS, ASSET, and ACCUPLACER

Trigonometry is a sine of the times.

—AUTHOR UNKNOWN

This chapter covers a lot of ground, and might take you longer than 20 minutes to complete. Split up your study into two parts if you need extra time to make your way through this lesson's examples.

LESSON SUMMARY

Sine, as mentioned in the quote, is one of the critical ratios that make up trigonometry basics. This lesson reviews each of the common ratios and how they are used to find missing sides and angles of right triangles. Special cases of right triangles are also explored, as well as angles of depression and elevation. The lesson concludes by explaining how each of the fundamental trigonometric functions is graphed, and how these graphs can be altered based on their equations.

Right Triangle Trigonometry

Right triangle trigonometry is based on the ratios between the lengths of the sides of a right triangle, which relate to the measures of the acute angles in the triangle. Let's start by reviewing the parts of a right triangle. Recall the parts of the right triangle identified in Lesson 11.

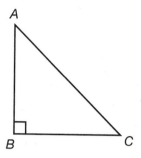

A

B

C

Take angle *A* in the right triangle *ABC*. The hypotenuse, or side across from the right angle, is side *AC*. The side adjacent angle *A* that is not the hypotenuse is side *AB*. The side opposite, or the farthest away from, angle *A* is side *BC*. The three basic trigonometric ratios, which are sine (sin), cosine (cos), and tangent (tan), can be found by using the appropriate sides of a right triangle such as the one shown, with respect to the indicated angle.

QUICK FACTS

The three basic trigonometric ratios can be summarized as sin $x \frac{opposite\ side}{hypotenuse}$, cos $x \frac{adjacent\ side}{hypotenuse}$, and tan $x = \frac{opposite\ side}{adjacent\ side}$. These ratios can be remembered by using the mnemonic device **SOH CAH TOA**.

Because the hypotenuse is always the longest side of any right triangle, the fractions for sine and cosine will always be proper, and therefore the sine and cosine values will always be between −1 and 1.

Here are some examples on how to use these ratios.

Examples

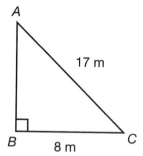

A

17 m

B   8 m   C

To find the sine of angle *A*, use the ratio sin *A* = $\frac{opposite\ side}{hypotenuse}$ and substitute the values from the triangle. Be sure to use the length of the side opposite from angle *A* as the numerator and the length of the hypotenuse as the denominator. In this particular triangle the side opposite angle *A* measures 8 m and the hypotenuse measures 17 m. Therefore, sin $A = \frac{opposite\ side}{hypotenuse} = \frac{8}{17} \approx 0.47059$.

The inverse of a trigonometric function is the reverse of finding the actual value of the function. It is used, in this case, to find the angle measure when the particular trigonometric value is known.

To find the measure of angle *A*, now use the inverse function or sin⁻¹*A* on a calculator. Be sure that your calculator is in the correct mode (degrees). Substitute the value of the sin for *A* in the inverse function to get sin⁻¹($\frac{8}{17}$) ≈ 28.0725. The measure of angle *A* is 28.1°, to the nearest tenth of a degree.

Trigonometric ratios can also be used to find a missing side of a right triangle. Take, again, the triangle in the figure. To find the missing side, which is adjacent to angle *A*, choose the cosine ratio cos *A* = $\frac{adjacent\ side}{hypotenuse}$. Because angle *A* is now known to be 28.1°, use that value for *A* and the given value of 17 m for the hypotenuse. Substitute the known values into the cosine ratio. Use *x* as the unknown side:

$$\cos 28.1 = \frac{x}{17}$$

Place cos 28.1 over 1 to form a proportion and find the cos 28.1 on a calculator:

$$\frac{\cos 28.1}{1} = \frac{x}{17}$$

$$\frac{0.8821}{1} = \frac{x}{17}$$

Find the cross products of the proportion to solve for x:

$$1x = 14.9962$$
$$x \approx 15.0 \text{ m}$$

DID YOU KNOW?

When all three sides are solved for in a right triangle, the Pythagorean theorem ($a^2 + b^2 = c^2$ from Lesson 11) can be used to check the result.

TIP

When the angle measure is known and you are looking for a missing side of a right triangle, use the function sin, cos, or tan. When the angle measure is unknown and the sides are given, use the inverse function $\sin^{-1}$, $\cos^{-1}$, or $\tan^{-1}$.

Practice

Using this figure, find the value of sin, cos, or tan as a decimal.

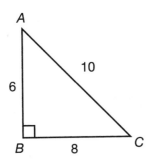

1. Find tan A

2. Find sin A

3. Find cos C

4. Find sin C

5. Find tan C

Use the appropriate trigonometric function to find the measure of angle A to the nearest degree in each triangle.

6.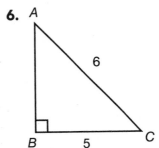

7.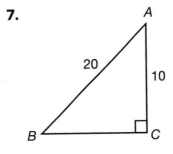

8.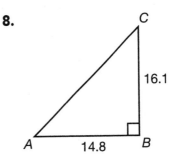

Special Right Triangles

There are a few types of special right triangles that are used frequently in trigonometry. Knowing the special relationships within these triangles is essential to help you to solve many problems.

45–45–90 Right Triangles

The first type of special right triangle is the 45–45–90 right triangle. In this triangle, the two acute angles have the same measure; therefore, the two legs must also have the same measure. This makes the triangle an isosceles right triangle. In this case, the sine and cosine of the acute angles will be the same, and the tangent of the acute angles will be equal to 1. In addition, the length of the hypotenuse will always be equal to the length of a leg multiplied by $\sqrt{2}$. You can use the Pythagorean theorem to see why that is true.

Example

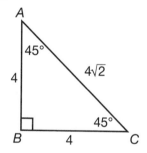

In this triangle, $\sin A = \frac{opposite\ side}{hypotenuse} = \frac{4}{4\sqrt{2}} = \frac{1}{\sqrt{2}} = \frac{\sqrt{2}}{2}$, $\cos A = \frac{adjacent\ side}{hypotenuse} = \frac{4}{4\sqrt{2}} = \frac{1}{\sqrt{2}} = \frac{\sqrt{2}}{2}$, and $\tan A = \frac{opposite\ side}{adjacent\ side} = \frac{4}{4} = 1$.

30–60–90 Right Triangles

Another type of special right triangle is the 30–60–90 right triangle. In this case, the smallest side across from the 30° angle is always half the length of the hypotenuse, and the side across from the 60° angle is always equal to the smallest side multiplied by $\sqrt{3}$. You can use the Pythagorean theorem to test this relationship as well.

Example

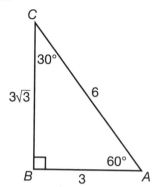

In this triangle, $\sin 60° = \frac{opposite\ side}{hypotenuse} = \frac{3\sqrt{3}}{6} = \frac{\sqrt{3}}{2}$, $\cos 60° = \frac{adjacent\ side}{hypotenuse} = \frac{3}{6} = \frac{1}{2}$, and $\tan 60° = \frac{opposite\ side}{adjacent\ side} = \frac{3\sqrt{3}}{3} = \sqrt{3}$.

QUICK FACT

When angles are complements, they add to 90 degrees. The cosine of an angle is equal to the sine of its complement. Here are some common examples.

$$\cos 60° = \sin 30° = \frac{1}{2} = 0.5$$
$$\sin 60° = \cos 30° = \frac{\sqrt{3}}{2} \approx 0.8660$$
$$\cos 45° = \sin 45° = \frac{\sqrt{2}}{2} \approx 0.7071$$

DID YOU KNOW?

When calculating sine, cosine, and tangent, the acute angles of the triangle are used, but not the right angle. The tan 90° is undefined because you cannot divide by zero. Recall that when cosine is equal to zero, the denominator of the tangent ratio will be zero.

Practice

Find the indicated value in each of these problems.

9.

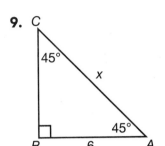

10.

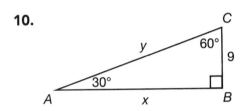

11.

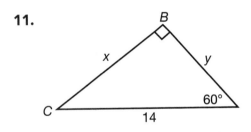

12. If cos 0° = 1, then sin 90° = _____.

13. If sin 40° = 0.6428, then cos _____° = 0.6428.

14. If cos 10° = 0.9848, then sin _____° = 0.9848.

Angles of Elevation and Depression

In trigonometric applications, two common terms that are used are the angle of elevation and the angle of depression.

DEFINITIONS

The *angle of elevation* is the angle formed by the horizontal line and the line that forms the direct distance to the top of an object when the object is elevated from where the viewer is located. The *angle of depression* is the angle formed by the horizontal line and the direct distance to the object when the object is below where the viewer is located. These are both illustrated in the following figures.

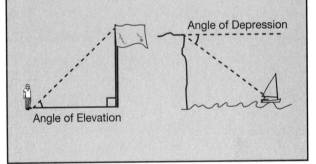

Angles of depression and elevation are congruent to each other, because they are alternate interior angles. In order to solve a problem involving these types of angles, first draw a picture and identify the parts of the triangle in order to select the correct trigonometric ratio.

Example

A person in a boat on the water is 200 yards away from the shore. He sights the top of a lighthouse there. If the angle of elevation is 25 degrees, how tall is the lighthouse?

A possible diagram of this situation is shown in the following figure. You should always draw a picture for this type of problem.

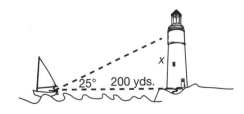

In this situation, the angle of 25° is the angle of elevation, the distance between the boat and the lighthouse is the adjacent side, and the height of the lighthouse is the side opposite the angle. Since the opposite and adjacent sides are involved, use tangent to solve this problem.

First, substitute into the formula. Use x as the height of the lighthouse:

$$\tan 25° = \frac{opposite\ side}{adjacent\ side} = \frac{x}{200}$$

Place tan 25° over 1 to form a proportion and find the tan 25 on a calculator.

$$\frac{\tan 25}{1} = \frac{x}{200}$$

$$\frac{0.4663}{1} = \frac{x}{200}$$

Find the cross products of the proportion to solve for x.

$$1x = 93.2615$$
$$x \approx 93.3 \text{ yds.}$$

Practice

Find the indicated value in each of these problems.

15. A person is on the rooftop of a building 65 feet tall and sights a person walking on the street below. If the angle of depression formed is 36 degrees, what is the direct distance between the two people to the nearest tenth of a foot?

16. A 50 m wire is attached to the top of a cellular tower and anchored to the ground 37 meters from the base of the tower. What is the angle of elevation formed by the ground and the wire?

17. An airplane is 1,260 feet above the ground, and the pilot sights a runway on the ground with an angle of depression of 24 degrees. What is the diagonal distance the plane is from the runway?

18. A person is 150 yards from the base of a cliff. If the cliff has a height of 220 yards, what is the angle of elevation from where the person is standing?

DID YOU KNOW?

Angle measure in trigonometry can be in degrees, as mentioned, or in radian measure. Radian measure can be converted to degrees using the fact that π radians are equal to 180°: $\frac{\pi}{2} = 90°$, $\frac{\pi}{3} = 60°$, $\frac{\pi}{4} = 45°$, and $\frac{\pi}{6} = 30°$.

Common Trigonometric Graphs

There are three types of common trigonometric graphs that you may encounter on college placement exams. They are the various forms of $y = \sin x$, $y = \cos x$, and $y = \tan x$. These functions are known as *periodic functions*, because they have a pattern that repeats within a certain interval. (Note: Your understanding of trigonometric concepts will be expanded in Lesson 19, as well.)

QUICK FACTS

In each of the following trigonometric functions, a represents the amplitude of the function, which is the highest and lowest the function will go. For example, with amplitude of 3, the graph will contain values between 3 and –3 on the y-axis. The value of b in the equation represents the frequency. In other words, if the frequency is 2 then there are two cycles of the function between 0 and 2π (360°).

y = a sin bx

In the graph of $y = \sin x$, $a = 1$ and $b = 1$. The amplitude is 1, so the graph oscillates between 1 and –1 on the y-axis. Because the frequency is also 1, then one cycle of the function is contained between 0 and 2π. Take a few known points on the sine function to sketch the graph. For example, $\sin 0° = 0$, $\sin 30° = \frac{1}{2}$, $\sin 60° = 0.8660$ or $\frac{\sqrt{3}}{2}$, $\sin 90° = 1$, $\sin 180° = 0$, $\sin 270° = -1$, and so on. Plotting these points on a graph and connecting with a smooth curve results in the following figure.

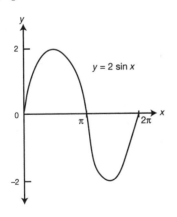

If the amplitude were to change, as in the graph of the function $y = 2 \sin x$, the graph would remain the same except for the fact that it would oscillate between 2 and –2 because $a = 2$. This is shown in the following graph.

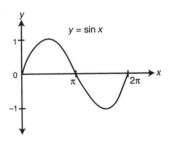

If the frequency were to change, as in the function $y = \sin 2x$, there would be two cycles of the graph between 0 and 2π. In this case, the amplitude would remain at 1. The graph of this function is shown here.

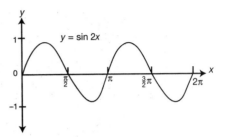

If the amplitude is a negative value, the y-values in the function become their opposites, and the figure is reflected in the x-axis as shown here.

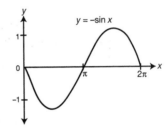

y = a cos bx

In the graph of $y = \cos x$, $a = 1$ and $b = 1$. The amplitude is 1, so the graph oscillates between 1 and –1 on the y-axis. Because the frequency is also 1, then one cycle of the function is contained between 0 and 2π. Take a few known points on the cosine function to sketch the graph. For example, $\cos 0° = 1$, $\cos 30° = 0.8660$ or $\frac{\sqrt{3}}{2}$, $\cos 60° = \frac{1}{2}$, $\cos 90° = 0$, $\cos 180° = -1$, $\cos 270° = 0$, and so on. Plotting these points on a graph and connecting with a smooth curve results in this figure:

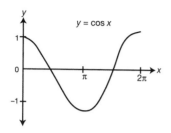

If the amplitude and frequency were to change, as in the function $y = 3 \cos 2x$, the amplitude is 3 and there would be two cycles of the graph between 0 and 2π. The graph of this function is shown here.

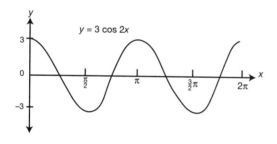

$y = \tan bx$

Remember that division by zero is undefined. Because $\tan x = \frac{\sin x}{\cos x}$ (a fact that will be covered in more detail in Lesson 19), and $\cos x = 0$ when $x = 90°\left(\frac{\pi}{2}\right)$ or $270°$ $\left(\frac{3\pi}{2}\right)$, then $\tan x$ is undefined at $90°\left(\frac{\pi}{2}\right)$ and $270°$ $\left(\frac{3\pi}{2}\right)$. At these places on the graph, there are vertical *asymptotes*. These are usually denoted as a dashed line that the graph may not cross. Take these facts and a few known points on the tangent function to sketch the graph. For example, $\tan 0° = 0$, $\tan 45° = 1$, $\tan 135° = -1$, $\tan 180° = 0$, $\tan 225° = 1$, $\tan 360° = 0$, and so on. Plotting these points on a graph with the vertical asymptotes and connecting with a smooth curve results in the following figure.

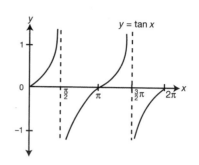

Practice

Match each of the following trigonometric graphs with the correct equation.

a. $y = 4 \cos \frac{1}{2}x$
b. $y = 3 \sin x$
c. $y = 3 \cos x$
d. $y = 4 \cos 2x$
e. $y = -\sin 2x$
f. $y = 3 \tan x$

19.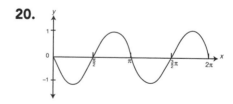

20.

21.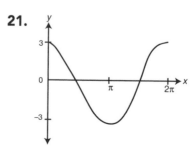

Practice Answers

22.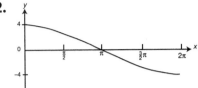

23.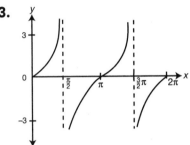

24.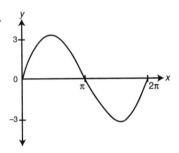

1. $1.\overline{3}$
2. 0.8
3. 0.8
4. 0.6
5. 0.75
6. $\sin A = \frac{5}{6}$; 56°
7. $\cos A = \frac{10}{20}$; 60°
8. $\tan A = \frac{16.1}{14.8}$; 47°
9. $6\sqrt{2}$
10. $x = 9\sqrt{3}$; $y = 18$
11. $x = 7\sqrt{3}$; $y = 7$
12. 1
13. 50
14. 80
15. 110.6 ft.
16. 42.3°
17. 3,097.8 ft.
18. 55.7°
19. d
20. e
21. c
22. a
23. f
24. b

19 ▶ TRIGONOMETRIC IDENTITIES AND EQUATIONS
For the COMPASS and ACCUPLACER

The mathematical sciences particularly exhibit order, symmetry, and limitation; and these are the greatest forms of the beautiful.
—ARISTOTLE

This chapter covers a lot of ground, and might take you longer than 20 minutes to complete. Split up your study into two parts if you need extra time to make your way through this lesson's examples.

LESSON SUMMARY
Lesson 19 reviews and expands on some of the important trigonometric concepts presented in the previous lesson. It begins with a review of the unit circle and its many implications, and then continues to the important trigonometric identities, and finally covers how to solve trigonometric equations.

The Unit Circle

Trigonometry can be expressed on the coordinate system. Because of the cyclic nature of these ratios, relationships emerge as points on a circle. This particular circle has a radius of one unit and center on the origin. This is known as the unit circle, and is shown in the following figure.

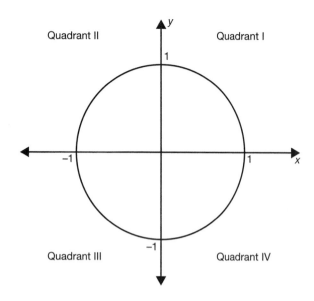

When an angle is formed between the *x*-axis and another ray with its vertex at the origin, the *x*- and *y*-coordinates of the point where the ray intercepts the circle are known as the cosine and sine of the angle, respectively. Take, for example, this figure:

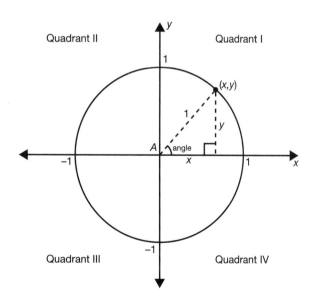

A right triangle is formed whose hypotenuse is a radius of the unit circle. One leg is along the *x*-axis, and the other is the vertical line segment connecting the point (*x*,*y*) to the *x*-axis. Because the unit circle

has a radius of 1, the cos $A = \frac{adjacent\ side}{1}$ and the sin $A = \frac{opposite\ side}{1}$. Thus, the *x*-coordinate of the point on the circle is the length of the adjacent side, the cosine of the angle, and the *y*-coordinate is the length of the opposite side, the sine of the angle.

Similarly, the tangent of angle *A* can be found using the ratio tan $A = \frac{opposite\ side}{adjacent\ side} = \frac{y}{x} = \frac{\sin}{\cos}$.

Positive and Negative Values

Knowing these relationships can be very helpful in determining when a sine, cosine, or tangent value is positive or negative. Recall this concept from Lesson 10. For example, where *x*-coordinates are negative (quadrants II and III), cosine values are also negative. Where *y*-coordinates are negative (quadrants III and IV), sine values are negative. Where *x*- and *y*-coordinates are opposite values (quadrants II and IV), tangent values are negative. This was apparent on the previous trigonometric graphs.

Values of sine are positive in the first and second quadrants, so the graph for sine was above the *x*-axis from 0 to π radians (0 to 180 degrees). Because sine is negative in the third and fourth quadrants, the graph for sine was below the *x*-axis from π to 2π radians (180 to 360 degrees).

Reference Angles

A *reference angle* is the acute angle from quadrant I that is used to find the trigonometric values of non-acute angles in other quadrants. For example, cosine is positive in quadrant I and quadrant IV, so an angle of 60º has the same cosine value as the angle 60º from the *x*-axis in quadrant IV. Thus, cos 60° = cos (360 − 60) = cos 300°. The reference angle is 60° and both cosines are equal to $\frac{1}{2}$. In the same way, sine is positive in quadrant I and quadrant II. Therefore, sin 45° is equal to the sine of the angle in the second quadrant that is 45 degrees from the *x*-axis. The reference angle is 45 degrees. This angle is equal to 180° − 45° = 135°. So, sin 45° = sin 135° = $\frac{\sqrt{2}}{2} \approx 0.7071$.

For an angle located in the third quadrant, such as 190°, the reference angle would be 190° − 180° = 10° for any of the trigonometric functions.

Example

If cos 45° = 0.7071, then the value of
cos 315° = _____.

To solve this problem, first identify the reference angle by subtracting 360° − 315° = 45°. Then, since cos 45° = 0.7071, and cosine is positive in the fourth quadrant, cos 315° = 0.7071.

TIP

A reference angle is always measured from the x-axis, not the y-axis, and will be an acute angle measure.

There are times when a trigonometric value may be given, along with a quadrant, and another trigonometric value will be asked for.

Example

If the cos $x = -\frac{3}{5}$, what is the value of tan x if it is located in quadrant II?

Recall from Lesson 18 that in right triangle trigonometry, cos $x = \frac{adjacent\ side}{hypotenuse}$.

Draw a picture of a right triangle, labeling the adjacent side to the indicated angle 3 units and the hypotenuse 5 units, as shown in this figure:

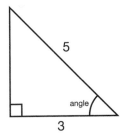

Use the Pythagorean theorem $a^2 + b^2 = c^2$; $a = 3$ and $c = 5$.

Substitute:
$$3^2 + b^2 = 5^2$$
Apply the exponents:
$$9 + b^2 = 2^5$$
Subtract 9 from each side of the equation:
$$9 - 9 + b^2 = 25 - 9$$
$$b^2 = 16$$
Take the square root of each side of the equation:
$$\sqrt{b^2} = \sqrt{16}, \text{ so } b = 4 \text{ units and is the oppo-}$$
site side from the angle.

Since tan $x = \frac{opposite\ side}{adjacent\ side}$, the value of tan $x = \frac{4}{3}$. It is located in quadrant II where tangent is negative, so the final answer is $-\frac{4}{3}$.

TIP

Draw a picture for any problem involving geometric shapes. The problem is usually easier if you can see the diagram.

Practice

Answer each of the following.

1. Name the quadrants in which each of the following functions is negative:
   a. sin = _____ and _____
   b. cos = _____ and _____
   c. tan = _____ and _____

2. Measure the reference angle if angle $A = 100°$.

3. Measure the reference angle if angle $A = 190°$.

4. Measure the reference angle if angle $A = 355°$.

5. If sin 60° = 0.8660, then the value of sin 240° = _____.

6. If tan 225° = 1, then the value of tan 45° = _____.

7. If cos 15° = 0.9659, then the value of cos 165° = _____.

8. If the cos $x = \frac{3}{5}$, what is the value of tan x if it is located in quadrant IV?

9. If the sin $x = \frac{8}{17}$, what is the value of cos x if it is located in quadrant III?

10. If the tan $x = \frac{5}{12}$, what is the value of cos x if it is located in quadrant I?

Trigonometric Identities

Identities are equations that hold true for any value. The basic trigonometric identities are:

$$\sin x = \frac{1}{\csc x} \quad \cos x = \frac{1}{\sec x}$$

$$\tan x = \frac{\sin x}{\cos x} \quad \cot x = \frac{\cos x}{\sin x}$$

Because they are identities, one side of each equation can be a replacement for the other side. Trigonometric identities can be used to simplify trigonometric expressions, and later in this lesson, will be used to help solve trigonometric equations.

Example
Express sin x cot x as a single term containing one function.

First, substitute using the basic identities. Because cot $x = \frac{\cos x}{\sin x}$, the expression becomes $\sin x \times \frac{\cos x}{\sin x}$. Cancel the common factors of sin x to get $\cancel{\sin x} \times \frac{\cos x}{\cancel{\sin x}} = \cos x$.

Practice
Express each of the following expressions as a single term containing one function.

11. sec x cot x

12. 3 cos x tan x

13. $\frac{\cos(-x)}{\sec x}$

14. csc x cos x tan$(-x)$

Pythagorean Identities

There are three important trigonometric identities based on the Pythagorean theorem, $a^2 + b^2 = c^2$.

They are:

$$\sin^2 x + \cos^2 x = 1$$
$$\tan^2 x + 1 = \sec^2 x$$
$$\cot^2 x + 1 = \csc^2 x$$

DID YOU KNOW?

Each of the Pythagorean identities can be altered using equation-solving rules.

$\sin^2 x + \cos^2 x = 1$ $\qquad\qquad$ $\sin^2 x = 1 - \cos^2 x$ $\qquad\qquad$ $\cos^2 x = 1 - \sin^2 x$

$\tan^2 x + 1 = \sec^2 x$ $\qquad\qquad$ $\tan^2 x + 1 = \sec^2 x$ $\qquad\qquad$ $\tan^2 x = \sec^2 x - 1$

$\cot^2 x + 1 = \csc^2 x$ $\qquad\qquad$ $1 + \cot^2 x = \csc^2 x$ $\qquad\qquad$ $\cot^2 x = \csc^2 x - 1$

These identities can also be used to simplify trigonometric expressions.

Example

Express $\sin^2 x - 1 + \cos x$ as a single term containing one function.

To simplify this expression, use the first Pythagorean identity ($\sin^2 x + \cos^2 x = 1$) and subtract $\cos^2 x$ and 1 from each side of the identity.

The identity becomes $\sin^2 x - 1 = -\cos^2 x$.

Now substitute into the expression to get the single function $\sin^2 x - 1 + \cos x = -\cos^2 x + \cos x$. The expression is now in terms of just one function, the cosine.

Practice

Express each of the following expressions as a single term containing one function.

15. $\cos x(\tan^2 x + 1)$

16. $\sin^2 x(1 + \cot^2 x)$

17. $\dfrac{1 - \sin^2 x}{\cos x}$

18. $\dfrac{\sin^2 x + \cos^2 x}{\sec x}$

Solving Trigonometric Equations

When solving trigonometric equations, you will use many of the steps that were used when solving linear and quadratic equations in earlier lessons. The trigonometric identities mentioned in the previous sections can also be substituted to help in the equation-solving process.

Example 1

Find all values satisfying the equation $2 \sin x = 1$ in the interval $0° \le x < 360°$.

Take the equation $2 \sin x = 1$, and divide each side by 2 to get $\sin x$ alone.

The equation becomes $\frac{2 \sin x}{2} = \frac{1}{2}$, so $\sin x = \frac{1}{2}$.

Because sin is positive in quadrants I and II, and $\sin 30° = \frac{1}{2} [\sin^{-1}(\frac{1}{2}) = 30°]$, then the values are $x = 30°$ and $x = 180° - 30° = 150°$.

This equation can be checked by substituting in the value(s) of x in the solution set. To do this, first substitute the angle measure in for the value of x in the equation.

First check the solution of 30°:

$2 \sin x = 1$ becomes $2 \sin (30°) = 1$.

Because $\sin 30° = 0.5$, then the equation becomes $2(0.5) = 1$, and $1 = 1$.

Next, check the solution of 150 degrees:

$2 \sin x = 1$ becomes $2 \sin (150°) = 1$.

Because $\sin 150° = 0.5$, then the equation becomes $2(0.5) = 1$, and $1 = 1$.

Each of the solutions is correct.

$\sin x + 2 = 0$ or $\sin x + 1 = 0$
$\sin x = -2$ or $\sin x = -1$
$\sin^{-1}(-2)$ has no solution or
$\sin^{-1}(-1) = 270°$
($\sin x$ cannot be > 1)

Example 2

Find all values satisfying the equation $2 \cos x \sin x + \sqrt{2} \sin x = 0$ in the interval $0° \leq x < 360°$.

In this case, first factor out the common factor of $\sin x$ to get the equation:

$$\sin x (2 \cos x + \sqrt{2}) = 0$$

Set each factor equal to zero:

$$\sin x = 0 \text{ or } 2 \cos x + \sqrt{2} = 0$$

Solve for x by using the inverse functions:

$$\sin x = 0 \text{ or } 2 \cos x = -\sqrt{2}$$
$$\sin^{-1}(0) = 0 \text{ or } \cos x = -\frac{\sqrt{2}}{2}$$

Using the reference angle $45°$ $(180 - 135 = 45)$ also yields a solution of $225°$ $(180 + 45 = 225)$, since cosine is also negative in the third quadrant. In addition, $\sin(180) = 0$.

So the values that satisfy the equation are $x = 0°, 135°, 180°, 225°$.

Example 3

Find all values satisfying the equation $\sin^2 x + 3 \sin x + 2 = 0$ in the interval $0° \leq x < 360°$.

Treat this equation as you would a quadratic equation, as explained in Lesson 15.

Temporarily replace the sine function with a variable, and factor the left side of the equation as if it were the expression $z^2 + 3z + 2$; then substitute the sine functions back into the factors. The factors are $(z + 2)(z + 1)$, so replacing the sine functions into the factored equation results in $(\sin x + 2)(\sin x + 1) = 0$.

Set each factor equal to zero and solve for x using the inverse functions.

Example 4

Find all values satisfying the equation $\sec^2 x - 3 = \tan x$ in the interval $0° \leq x < 360°$.

First, get the equation in standard form:

$$\sec^2 x - \tan x - 3 = 0$$

Substitute $\tan^2 x + 1$ for $\sec^2 x$ from the identities to get the equation in terms of one function. The equation becomes $(\tan^2 x + 1) - \tan x - 3 = 0$.

This simplifies to $\tan^2 x - \tan x - 2 = 0$.

Next, factor the equation without $\tan x$ as you would factor $z^2 - z - 2 = (z - 2)(z + 1)$.

Replace the tangent function into the factors: $(\tan x - 2)(\tan x + 1) = 0$.

Set each factor equal to zero and solve for x using the inverse functions.

$$\tan x - 2 = 0 \quad \text{or} \quad \tan x + 1 = 0$$
$$\tan x = 2 \quad \text{or} \quad \tan x = -1$$
$$\tan^{-1}(2) \approx 63.435°, \quad \text{or} \quad \tan^{-1}(-1) = 135°,$$
$$243.435° \quad\quad\quad\quad 315°$$

Remember that each equation can be checked using the solutions. Use this as a strategy to help with multiple-choice questions.

Practice

Find all values satisfying the following equations in the interval $0° \leq x < 360°$.

19. $2 \cos x = \sqrt{3}$

20. $\tan x - 1 = 0$

21. $-2 \sin x = \sqrt{2}$

22. $2 \cos x \sin x - \cos x = 0$

23. $3 \tan x = -\sqrt{3}$

24. $\cos^2 x + 2 \cos x - 3 = 0$

25. $3 \sin x = 2 \cos^2 x - 3$

Practice Answers

1. a. III and IV
   b. II and III
   c. II and IV

2. $80°$

3. $10°$

4. $5°$

5. -0.8660

6. 1

7. -0.9659

8. $-\frac{4}{3}$

9. $-\frac{15}{17}$

10. $\frac{12}{13}$

11. $\frac{1}{\sin x} = \csc x$

12. $3 \sin x$

13. $\cos^2 x$

14. -1

15. $\frac{1}{\cos x} = \sec x$

16. 1

17. $\cos x$

18. $\frac{1}{\sec x} = \cos x$

19. $30°, 330°$

20. $45°, 225°$

21. $225°, 315°$

22. $30°, 90°, 150°, 270°$

23. $150°, 330°$

24. $0°$

25. $210°, 270°, 330°$

LESSON

20 ▶ COMPLEX NUMBERS, SEQUENCES AND SERIES, AND MATRICES
For the COMPASS, ASSET, and ACCUPLACER

If people do not believe that mathematics is simple, it is only because they do not realize how complicated life is.
—JOHN LOUIS VON NEUMANN

This chapter covers a lot of ground, and might take you longer than 20 minutes to complete. Split up your study into three parts if you need extra time to make your way through this lesson's examples. You can tackle complex numbers first, then sequences and series, and then matrices.

LESSON SUMMARY
If your goal is to place into an advanced mathematics course in college, you will need to be proficient on many advanced topics. Complex numbers are mathematical numbers that are not in the real number system. One part of a complex number is called an imaginary number. Sequences and series are lists of numbers with specific relationships to one another; series are sums based on sequences. Matrices are two-dimensional arrays of numbers with special characteristics and with defined operations. All of these topics have applications in physics and engineering.

Complex Numbers

Lesson 1 reviewed the types of real numbers in mathematics. This lesson reviews the complex numbers, which have two parts, a real component and an imaginary component.

Imaginary Numbers

These numbers were named *imaginary* in the seventeenth century because they were thought to have no useful purpose. It turns out that they are used to solve problems associated with electricity and physics. Consider the equation $x^2 + 1 = 0$. To solve for x, you subtract 1 from each side to get $x^2 = -1$. There is no real number that when squared yields a negative number. By taking the square root of both sides you get $x^2 = \pm\sqrt{-1}$. To give this expression meaning, the symbol i is defined to be $\sqrt{-1}$ and $-i = -\sqrt{-1}$. This is the base for the imaginary number system. If you encounter a negative sign within a square root symbol, you can separate it out as i. For example, $\sqrt{-4} = \sqrt{-1} \times \sqrt{4} = 2i$ and $\sqrt{-49} = 7i$. Note that the coefficient is shown before the i. If there is a real number component left over that is irrational, it is shown last in the notation. For example, $\sqrt{-75} = \sqrt{-1} \times \sqrt{25} \times \sqrt{3} = 5i\sqrt{3}$. Remember that the radical symbol is a grouping symbol in the order of operations, so evaluate the radical before any multiplication or division. For example, $-\frac{1}{3}\sqrt{-9} = -\frac{1}{3} \times 3i = -1i = -i$.

DEFINITION

The basis of the imaginary number system is $i = \sqrt{-1}$ and $-i = -\sqrt{-1}$.

Practice

Express each number in terms of i and in simplest form.

1. $\sqrt{-64}$

2. $\frac{1}{2}\sqrt{-100}$

3. $\frac{1}{4}\sqrt{-\frac{1}{4}}$

4. $-\sqrt{-32}$

Powers of i

Since i is defined to be $\sqrt{-1}$, $i^2 = (\sqrt{-1})(\sqrt{-1})$. Using the Laws of Exponents, it then follows that $i^3 = i^2 \times i = -i$ and $i^4 = i^2 \times i^2 = -1 \times -1 = 1$. Study the following table for the first 16 powers of i:

REMAINDER OF 1	REMAINDER OF 2	REMAINDER OF 3	REMAINDER OF 0
$i^1 = i$	$i^2 = -1$	$i^3 = -i$	$i^4 = 1$
$i^5 = i$	$i^6 = -1$	$i^7 = -i$	$i^8 = 1$
$i^9 = i$	$i^{10} = -1$	$i^{11} = -i$	$i^{12} = 1$
$i^{13} = i$	$i^{14} = -1$	$i^{15} = -i$	$i^{16} = 1$

Notice the relationship of the powers as they are grouped in columns. Each column results in a different remainder when the powers are divided by 4. Any power of i will be one of the four values as shown in the table. To evaluate the powers of i, divide the power of i by 4 and take note of the remainder. Then just evaluate the power of i using the remainder as the power. For example, if you divide $27 \div 4 = 6$ R 3. So $i^{27} = i^3 = -i$.

QUICK FACTS

When you divide a power of i by 4, the results are as follows. Let x be any whole number.

Remainder of 0	Remainder of 1	Remainder of 2	Remainder of 3
$i^{4x} = 1$	$i^{4x+1} = i$	$i^{4x+2} = -1$	$i^{4x+3} = -i$

Practice

Simplify:

5. i^{11}

6. i^{98}

7. i^{64}

8. $i^{2,013}$

Operations and Imaginary Numbers

You can add, subtract, multiply, and divide imaginary numbers, as long as you put the numbers in terms of i before performing the operations, just as you would when operating with real number radicals. To add $\sqrt{-25} + \sqrt{-121}$, convert to i form and then add as like terms: $5i + 11i = 16i$. Use a similar method for subtraction. For example, $\sqrt{-81} - \sqrt{-4} = 9i - 2i = 7i$. When you multiply and divide, again remember to first convert in terms of i and then multiply and divide the i portion.

TIP

Remember that when you perform operations with imaginary numbers you first convert the numbers to be in terms of i.

Example

For example, multiply $\sqrt{-25} \times \sqrt{-9} = 5i \times 3i = (5 \times 3)(i \times i) = 15i^2 = 15(-1) = -15$. You can multiply when the positive portions of the radicands are irrational. $\sqrt{-3} \times \sqrt{-48} = i\sqrt{3} \times 4i\sqrt{3} = (\sqrt{3})(4\sqrt{3})(i^2) = (4 \times 3)(-1) = -12$. Similarly, in division: $\sqrt{-100} \div \sqrt{-4} = \frac{10i}{2i} = 5$. Notice in division that the i portions of the numerator and denominator cancel.

QUICK FACTS

When you add or subtract imaginary numbers, the result is an imaginary number, except when the result is zero.

When you multiply or divide two imaginary numbers, the result is a real number.

Practice

Solve:

9. $\sqrt{-49} + \sqrt{-144}$

10. $\sqrt{-36} - \sqrt{-81}$

11. $\sqrt{-100} \times \sqrt{-9}$

12. $\sqrt{-2} \times \sqrt{-12}$

13. $\sqrt{-144} \div \sqrt{-16}$

14. $\frac{\sqrt{-48}}{\sqrt{-16}}$

Addition and Subtraction— Complex Numbers

DEFINITION

A *complex number* is a binomial (a number with two terms) of the form $a + bi$, where a is the real number term and bi is the imaginary term. Both a and b are real numbers.

You can treat addition and subtraction of complex numbers the same as when you combine like terms with polynomials. You combine the real terms, and separately combine the imaginary terms. $(4 + \sqrt{-16})$

+ $(-2 + \sqrt{-49}) = (4 + -2) + (4i + 7i) = 2 + 11i$. When subtracting, do not forget to distribute the coefficient -1.

Example

For example, change subtraction to addition by adding the opposite in the following problem: $(-10 - 12i) - (7 - 8i) = (-10 - 12i) + (-7 + 8i) = (-10 + -7) + (-12i + 8i) = -17 + -4i = -17 - 4i$.

Practice

Find the sum or difference:

15. $(6 - 5i) + (-6 + 7i)$

16. $(-3 + \sqrt{-40}) + (8 - \sqrt{-90})$

17. $(6 - 5i) - (-6 + 7i)$

18. $(-8 + 14i) - (10 - 6i)$

Multiplication of Complex Numbers

To multiply a complex number by another complex number, use the distributive property as you reviewed when multiplying two binomials in Lesson 14. To multiply $(4 + 3i)(2 + i)$, recall from Lesson 14 the acronym **FOIL** (**f**irst, **o**uter, **i**nner, **l**ast). Multiply, simplify the powers of i, and combine like terms.

> Multiply the first terms of each complex number. $4 \cdot 2 = 8$
> Multiply the outer terms. $4 \cdot i = 4i$
> Multiply the inner terms. $3i \cdot 2 = 6i$
> Multiply the last terms of each binomial.
> $3i \cdot i = 3i^2 = 3(-1) = -3$
> Combine the like terms: $(8 + -3) + (4i + 6i) = 5 + 10i$

Almost every multiplication of complex numbers results in another complex number. The only exception is when you multiply a complex number and its conjugate.

Example

To multiply $(10 + 3i)(10 - 3i)$:

> Multiply the first terms of each complex number. $10 \cdot 10 = 100$
> Multiply the outer terms. $10 \cdot -3i = -30i$
> Multiply the inner terms. $3i \cdot 10 = 30i$
> Multiply the last terms of each binomial.
> $3i \cdot -3i = -9i^2 = -9(-1) = 9$
> Combine the like terms: $(100 + 9) + (-30i + 30i) = 109 + 0i = 109$

Practice

Find the conjugate.

19. $12 + 8i$

20. $-5 - 3i$

Multiply and simplify.

21. $(4 + 6i)(2 + 7i)$

22. $(10 + 9i)(2 - 4i)$

23. $(8 + 10i)(8 - 10i)$

Dividing Complex Numbers

When you divide complex numbers, rewrite the problem as a fraction, and multiply both the numerator and the denominator by a form of one that uses the conjugate of the denominator. This is done so that the denominator of the quotient will be a real number, said to be *rationalized*.

Example

For example, to divide $(3 + 2i) \div (2 + 4i)$, rewrite and multiply as described before: $\frac{3 + 2i}{2 + 4i}$ $\times \frac{2 - 4i}{2 - 4i} = \frac{6 + -12i + 4i - 8i^2}{4 + 16} = \frac{14 - 8i}{20}$. This can be simplified by dividing out the common factor of 2 from each term and writing in complex number form: $\frac{14 - 8i}{20} = \frac{7 - 4i}{10} = \frac{7}{10} - \frac{2}{5}i$.

Practice

Divide and simplify.

24. $(4 - 3i) \div (2 + 2i)$

25. $\frac{1 + 2i}{4 - i}$

Sequences

DEFINITION

A *sequence* is an ordered set of numbers. Each number is called a term or an element. A sequence is often referred to as A, and an element is referred to as a_n, where n, called the index, names a specific element in the sequence.

An example of a sequence is $A = \{1, 4, 7, 10, 13, 16, \ldots\}$. There are six elements shown in this example; $a_1 = 1$, $a_2 = 4$, and so on. Often, sequences have a pattern. This sequence has the pattern where you add 3 to an element to get the next element in the sequence. This sequence can be shown as a function rule. Because the difference is 3, the rule is $3n - 2$, where n is the element number. Thus you can find the next element not shown, so $a_6 = 16$, $a_7 = 19$, and $a_8 = 22$.

There are two common types of sequences, arithmetic and geometric.

Arithmetic Sequences

The preceding sequence is an example of an arithmetic sequence. An *arithmetic sequence* is one in which consecutive elements have a constant difference. The number added to an element to get the next element in the sequence is called the *common difference*, and is usually denoted by d. The common difference d in the sequence $A_n = \{-11, -7, -3, 1, 5, 9, \ldots\}$ is 4:

$$9 - 5 = 4$$
$$5 - 1 = 4$$
$$1 - -3 = 4$$
$$-3 - -7 = 4$$
$$-7 - -11 = 4$$

The common difference can also be negative, such as in $A_n = \{8, 5, 2, -1, -4, -7, \ldots\}$:

$$-7 - -4 = -3$$
$$-4 - -1 = -3$$
$$-1 - 2 = -3$$
$$2 - 5 = -3$$
$$5 - 8 = -3$$

The common difference, d, is -3.

Because the common difference involves repeated addition, there is a formula for finding an element in a sequence, a_n, when you know the difference and the value of the first element. So the value of the nth term in an arithmetic sequence where a (i.e., a_1) represents the first element of the sequence is $a_n = a + (n - 1)d$.

Example

For example, to find the eighth element, a_8 in the sequence $A_n = \{5, 9, 13, 17, 21, \ldots\}$ note that the common difference, d, is 4. So $a_8 = 5 + (8 - 1)4 = 5 + 7 \times 4 = 33$.

Geometric Sequences

A *geometric sequence* is one in which consecutive elements have a common ratio. The number multiplied by each element to get the next element in the sequence is called the common ratio, and is usually denoted by r. The common ratio r in $A_n = \{\frac{1}{9}, \frac{1}{3}, 1, 3, 9, 27, \ldots\}$ is 3:

$$27 \div 9 = 3$$
$$9 \div 3 = 3$$
$$3 \div 1 = 3$$
$$1 \div \frac{1}{3} = 3$$
$$\frac{1}{3} \div \frac{1}{9} = 3$$

If the ratio is negative, then note how the terms alternate signs. The common ratio r in $A_n = (8, -4, 2, -1, \frac{1}{2}, -\frac{1}{4}, \ldots)$ is $-\frac{1}{2}$:

$$-\frac{1}{4} \div \frac{1}{2} = -\frac{1}{2}$$
$$\frac{1}{2} \div -1 = -\frac{1}{2}$$
$$-1 \div 2 = -\frac{1}{2}$$
$$2 \div -4 = -\frac{1}{2}$$
$$-4 \div 8 = -\frac{1}{2}$$

Just as for arithmetic sequences, there is a formula for finding an element in the sequence a_n, when you know the ratio and the value of the first element. So the value of the nth term in a geometric sequence where a (i.e., a_1) represents the first element of the sequence is $a_n = ar^{(n-1)}$.

Example

For example, to find the seventh element, a_7, in the sequence $A_n = \{4, 12, 36, 108, 324, \ldots\}$, note that the common ratio, r, is 3. So $a_7 = 4(3)^6 = 4 \times 729 = 2{,}916$.

Practice

Tell whether the following sequences are arithmetic or geometric. Then find the common difference or ratio, and the eighth element in the sequence.

26. $A_n = \{-3, -1, 1, 3, 5, \ldots\}$

27. $A_n = \{32, 16, 8, 4, 2, \ldots\}$

28. $A_n = \{22, 16, 10, 4, -2, \ldots\}$

29. $A_n = \{2, -6, 18, -54, 162, \ldots\}$

Series

Arithmetic Series

DEFINITION

An *arithmetic series* is the sum of a finite number of elements in an arithmetic sequence.

The formula to find the sum of the first n terms of an arithmetic sequence, starting with the first element, is $\sum_{i=1}^{n} a_i = (\frac{n}{2})(a_1 + a_n)$.

In the preceding formula, note that a_n is the final element in the partial sum. The sum is the average of the smallest and the largest, multiplied by the number of elements. Given the arithmetic sequence $A_n = \{2, 6, 10, 14, 18, \ldots\}$, you can find the sum of the elements from the first to the tenth.

Example

The tenth element is given by $a_n = a + (n-1)d = 2 + 9(4) = 38$. So the partial sum is $\sum_{i=1}^{10} a_i = (\frac{10}{2})(2 + 38) = 5 \times 40 = 200$. The sequence can be represented with the function rule $4n - 2$. The multiplier is 4 because that is the common difference, and it is minus 2 by trial and error. This series could have been represented in a problem as:

Find $\sum_{n=1}^{10} 4n - 2$. You can use the rule to find the first element, $n = 1$, and the last element, $n = 10$, and then use the formula to find the sum as demonstrated earlier.

Geometric Series

DEFINITION

A *geometric series* is the sum of a finite number of elements in a geometric sequence.

The formula to find the sum of the first n terms of a geometric sequence, starting with the first element, is $\sum_{i=1}^{n} a_i = a(\frac{1-r^n}{1-r})$, where r is the common ratio.

Example

Given the geometric sequence $A_n = \{9, 27, 81, 243, \ldots\}$ you can find the sum of the first eight elements. Note that the common ratio is 3. Use the formula, and the sum of the first eight elements is $\sum_{n=1}^{8} a_n = 9(\frac{1-3^8}{1-3}) = \frac{3x - 6{,}560}{-2} = \frac{-19{,}680}{-2} =$ 29,520.

Practice

Find the partial sums.

30. $A_n = \{6, 2, -2, -6, -10, \ldots\} \sum_{n=1}^{10} a_n$

31. $A_n = \{2, -4, 8, -16, 32, \ldots\} \sum_{n=1}^{8} a_n$

32. $A_n = \{1, 6, 11, 16, 21, \ldots\} \sum_{n=1}^{12} a_n$

33. $\sum_{n=1}^{15} 3n - 2$

Matrices

DEFINITIONS

A *matrix* is a rectangular arrangement of elements arranged in rows and columns. An $m \times n$ matrix is said to have m rows and n columns. An example is shown here of a 2×3 matrix, which has two rows and three columns.

$$\begin{bmatrix} 4 & 5 & 14 \\ 3 & 6 & 15 \end{bmatrix}$$

You can refer to elements in the matrix with the letter a and two subscripts; the first subscript denotes the row and the second denotes the column.

$$\begin{bmatrix} a_{11} & a_{12} & a_{13} \\ a_{21} & a_{22} & a_{23} \end{bmatrix}$$

Matrix Operations

You can perform the operations of addition, subtraction, and multiplication on matrices. There are certain conditions that must be met, and there are specific rules to follow, which are explained next.

Addition and Subtraction

You can add or subtract only matrices that have exactly the same dimensions. When you have two matrices with the same dimensions, you simply add or subtract corresponding elements. The answer matrix will be these sums or differences in a matrix of the same dimensions.

Example

To add matrix **A** to **B**

where $\mathbf{A} = \begin{bmatrix} -4 & 3 \\ 5 & -6 \\ 7 & -1 \end{bmatrix}$ and $\mathbf{B} = \begin{bmatrix} 1 & 3 \\ 6 & 4 \\ -2 & -6 \end{bmatrix}$,

add each corresponding (positional) elements and put the answer in another 3×2 matrix:

$$\mathbf{A} + \mathbf{B} = \begin{bmatrix} -4+1 & 3+3 \\ 5+6 & -6+4 \\ 7+-2 & -1+-6 \end{bmatrix} = \begin{bmatrix} -3 & 6 \\ 11 & -2 \\ 5 & -7 \end{bmatrix}.$$

You cannot add $\begin{bmatrix} 1 & 2 & 3 \\ 4 & 5 & 6 \end{bmatrix}$ to $\begin{bmatrix} 5 & 7 \\ 3 & 3 \\ 8 & 2 \end{bmatrix}$

because the dimensions are not the same.

Subtraction works the same way. If matrix

$\mathbf{A}$ is $\begin{bmatrix} 3 & -5 & 7 \\ -9 & 3 & 4 \\ 2 & -2 & 1 \end{bmatrix}$ and $\mathbf{B}$ is $\begin{bmatrix} 1 & 5 & 2 \\ 6 & 3 & -7 \\ 4 & -6 & 5 \end{bmatrix}$,

then $\mathbf{A} - \mathbf{B}$ is $\begin{bmatrix} 3-1 & -5-5 & 7-2 \\ -9-6 & 3-3 & 4--7 \\ 2-4 & -2--6 & 1-5 \end{bmatrix} = \begin{bmatrix} 2 & -10 & 5 \\ -15 & 0 & 11 \\ -2 & 4 & -4 \end{bmatrix}.$

Scalar Multiplication

In scalar multiplication, you multiply a constant by a matrix. This is very straightforward; just multiply each element by the constant.

Example

For example, given matrix $\mathbf{A} = \begin{bmatrix} 2 & -4 \\ 6 & -8 \\ -10 & 5 \end{bmatrix}$

then $5\mathbf{A} = \begin{bmatrix} 5 \times 2 & 5 \times -4 \\ 5 \times 6 & 5 \times -8 \\ 5 \times -10 & 5 \times 5 \end{bmatrix} = \begin{bmatrix} 10 & -20 \\ 30 & -40 \\ -50 & 25 \end{bmatrix}.$

Similarly, $-2\mathbf{A} = \begin{bmatrix} -2 \times 2 & -2 \times -4 \\ -2 \times 6 & -2 \times -8 \\ -2 \times -10 & -2 \times 5 \end{bmatrix} = \begin{bmatrix} -4 & 8 \\ -12 & 16 \\ 20 & -10 \end{bmatrix}.$

Practice

Given that $\mathbf{A} = \begin{bmatrix} 1 & -5 & 5 \\ 6 & 3 & -2 \end{bmatrix}$ and $\mathbf{B} = \begin{bmatrix} 2 & 0 & 4 \\ -5 & 8 & -1 \end{bmatrix}$, evaluate.

34. $2\mathbf{A}$

35. $3\mathbf{B}$

36. $\mathbf{A} + \mathbf{B}$

37. $2\mathbf{A} - 3\mathbf{B}$

Matrix Multiplication

Matrix multiplication is not straightforward, and involves a complicated process. First, understand that the number of columns in the first matrix must equal the number of rows in the second matrix. The answer matrix will have dimensions that are the number of rows in the first matrix and the number of columns in the second matrix. For example, a 2×3 matrix can be multiplied by a 3×7 matrix, and will result in a 2×7 matrix. However, a 2×3 matrix cannot be multiplied by a 5×3 matrix, because the number of columns in the first matrix does not match the number of rows in the second. Generally, each row of the first matrix is multiplied by each column in the second matrix, matching subscripts. Each row's subproducts are added together. It is easiest to show symbolically:

$\begin{bmatrix} a & b & c \\ x & y & z \end{bmatrix} \times \begin{bmatrix} D \\ E \\ F \end{bmatrix} = \begin{bmatrix} aD + bE + cF \\ xD + yE + zF \end{bmatrix}$. Note that the answer matrix has two rows and one column. Another example shown symbolically:

$\begin{bmatrix} a & b & c \\ x & y & z \end{bmatrix} \times \begin{bmatrix} D & G \\ E & H \\ F & I \end{bmatrix} = \begin{bmatrix} aD + bE + cF & aG + bH + cI \\ xD + yE + zF & xG + yH = zI \end{bmatrix}.$

This answer matrix has dimensions of 2×2. Performing matrix multiplication requires meticulous attention to detail; double checking your work is a must.

Example

For example, if $\mathbf{A} = \begin{bmatrix} 4 & -2 \\ 6 & 4 \\ -1 & 5 \end{bmatrix}$ and $\mathbf{B} = \begin{bmatrix} 8 \\ -3 \end{bmatrix}$

then $\mathbf{A} \times \mathbf{B} = \begin{bmatrix} 4 \times 8 + -2 \times -3 \\ 6 \times 8 + 4 \times -3 \\ -1 \times 8 + 5 \times -3 \end{bmatrix} = \begin{bmatrix} 38 \\ 36 \\ -23 \end{bmatrix}.$

Practice

Given that $\mathbf{A} = \begin{bmatrix} 1 & -3 & 5 \\ 2 & -1 & -4 \end{bmatrix}$ and $\mathbf{B} = \begin{bmatrix} 2 \\ 4 \\ 6 \end{bmatrix}$ and

$\mathbf{C} = \begin{bmatrix} 5 & 0 & 7 \\ -2 & 4 & -1 \end{bmatrix}$ and $\mathbf{D} = \begin{bmatrix} 3 & -5 \\ 6 & -2 \\ 4 & 0 \end{bmatrix}$, evaluate.

38. $\mathbf{A} + \mathbf{C}$

39. $\mathbf{A} - \mathbf{C}$

40. $\mathbf{C} - \mathbf{A}$

41. $A \times B$

42. $C \times D$

Matrix Equations

Because you can add and subtract matrices with the same dimensions, you can solve equations involving matrices.

Examples
For example, you can solve for the matrix **C**, when $C + \begin{bmatrix} -2 & 3 \\ 4 & 6 \\ -1 & 0 \end{bmatrix} = \begin{bmatrix} 8 & 2 \\ 4 & 7 \\ 5 & 4 \end{bmatrix}$, by subtraction:

$C = \begin{bmatrix} 8 & 2 \\ 4 & 7 \\ 5 & 4 \end{bmatrix} - \begin{bmatrix} -2 & 3 \\ 4 & 6 \\ -1 & 0 \end{bmatrix}$ and so $C = \begin{bmatrix} 10 & -1 \\ 0 & 1 \\ 6 & 4 \end{bmatrix}$.

Another example is to solve for matrix **X**, when $3X - \begin{bmatrix} 2 & 5 \\ 10 & -5 \end{bmatrix} = \begin{bmatrix} 4 & 4 \\ 2 & 5 \end{bmatrix}$. First add the matrix to each side as shown next, and then divide each element of the resulting matrix by 3. $3X = \begin{bmatrix} 4 & 4 \\ 2 & 5 \end{bmatrix} + \begin{bmatrix} 2 & 5 \\ 10 & -5 \end{bmatrix}$ and therefore $X = \begin{bmatrix} 6 \div 3 & 9 \div 3 \\ 12 \div 3 & 0 \div 3 \end{bmatrix} = \begin{bmatrix} 2 & 3 \\ 4 & 0 \end{bmatrix}$.

Practice
Solve for the matrix **C**.

43. $C - \begin{bmatrix} 2 & 4 & -5 \\ -3 & 6 & 2 \end{bmatrix} = \begin{bmatrix} 5 & -7 & 9 \\ 2 & -4 & 1 \end{bmatrix}$

44. $2C + \begin{bmatrix} 1 & 3 \\ 2 & 5 \end{bmatrix} = \begin{bmatrix} 9 & 15 \\ 6 & 11 \end{bmatrix}$

Determinants

The *determinant* of a square matrix is a single value, either positive or negative, arrived at by a specific process involving multiplication, addition, and subtraction. The determinant of a matrix A is labeled as either $|A|$ or $\det(A)$. Determinants are calculated on square matrices.

Using symbols, $\det(A)$ where $A = \begin{bmatrix} w & x \\ y & z \end{bmatrix}$ is $wz - yx$. Think of the wz pair as a "down diagonal" and the yx pair as the "up diagonal." Notice that you multiply down the diagonal and then subtract the up diagonal from the down diagonal.

Example
For example, when $B = \begin{bmatrix} 5 & 3 \\ -2 & 4 \end{bmatrix}$, then the $\det(B)$, or labeled alternatively $\begin{vmatrix} 5 & 3 \\ -2 & 4 \end{vmatrix}$, is $5 \times 4 - -2 \times 3 = 20 + 6 = 26$.

Take note that while the symbol for determinant looks like the absolute value symbol, the determinant is *not* absolute value; a determinant can be negative.

Calculating the determinant of a 3×3 matrix is more complicated. There are several methods, but one method is to use the same idea that the determinant is found by subtracting the up diagonal from the down diagonal. The extra step is to duplicate the first two columns to then make a 3×5 matrix, and take the diagonals that have three elements.

Example
For example, to find the determinant of $\begin{bmatrix} 3 & -2 & 4 \\ 0 & 2 & 1 \\ 5 & -3 & 4 \end{bmatrix}$, duplicate the first two columns $\begin{bmatrix} 3 & -2 & 4 & 3 & -2 \\ 0 & 2 & 1 & 0 & 2 \\ 5 & -3 & 4 & 5 & -3 \end{bmatrix}$. Determine the down

diagonals $\begin{bmatrix} 3 & -2 & 4 & 3 & -2 \\ 0 & 2 & 1 & 0 & 2 \\ 5 & -3 & 4 & 5 & -3 \end{bmatrix}$, which are

$(3 \times 2 \times 4) + (-2 \times 1 \times 5) + (4 \times 0 \times -3) = 24 + -10 + 0 = 14$. Now, determine the up

diagonals $\begin{bmatrix} 3 & -2 & 4 & 3 & -2 \\ 0 & 2 & 1 & 0 & 2 \\ 5 & -3 & 4 & 5 & -3 \end{bmatrix}$, which are

$(5 \times 2 \times 4) + (-3 \times 1 \times 3) + (4 \times 0 \times -2) = 40 + -9 + 0 = 31$. Subtract to evaluate the determinant as $14 - 31 = -17$.

Practice

Evaluate the determinants of the matrices.

45. $\begin{bmatrix} 6 & -8 \\ 3 & -2 \end{bmatrix}$

46. $\det\begin{bmatrix} -4 & 7 \\ -1 & -5 \end{bmatrix}$

47. $\begin{bmatrix} 2 & -1 & 5 \\ 3 & -4 & 8 \\ 6 & 3 & 2 \end{bmatrix}$

48. $\det\begin{bmatrix} 2 & 4 & -6 \\ -1 & 3 & 3 \\ 4 & -2 & 5 \end{bmatrix}$

Practice Answers

1. $8i$

2. $5i$

3. $\frac{1}{8}i$

4. $-4i\sqrt{2}$

5. $-i$

6. -1

7. 1

8. i

9. $19i$

10. $-3i$

11. -30

12. $2\sqrt{6}$

13. 3

14. $\sqrt{3}$

15. $2i$

16. $5 - i\sqrt{10}$

17. $12 - 12i$

18. $-18 + 20i$

19. $12 - 8i$

20. $-5 + 3i$

21. $-34 + 40i$

22. $56 - 22i$

23. 164

24. $\frac{1}{4} - \frac{7}{4}i$

25. $\frac{2}{17} + \frac{9}{17}i$

26. Arithmetic; $a_8 = 11$; common difference is 2.

27. Geometric; $a_8 = \frac{1}{4}$; common ratio is $\frac{1}{2}$.

28. Arithmetic; $a_8 = -20$; common difference is –6.

29. Geometric; $a_8 = -4{,}374$; common ratio is –3.

30. –120; arithmetic sequence with a common difference of –4

31. –170; geometric sequence with a common ratio of –2

32. 342; arithmetic sequence with a common difference of 5

33. 330; arithmetic sequence with a common difference of 3, and $a = 1$

34. $\begin{bmatrix} 2 & -10 & 10 \\ 12 & 6 & -4 \end{bmatrix}$

35. $\begin{bmatrix} 6 & 0 & 12 \\ -15 & 24 & -3 \end{bmatrix}$

36. $\begin{bmatrix} 3 & -5 & 9 \\ 1 & 11 & -3 \end{bmatrix}$

37. $\begin{bmatrix} -4 & -10 & -2 \\ 27 & -18 & -1 \end{bmatrix}$

38. $\begin{bmatrix} 6 & -3 & 12 \\ 0 & 3 & -5 \end{bmatrix}$

39. $\begin{bmatrix} -4 & -3 & -2 \\ 4 & -5 & -3 \end{bmatrix}$

40. $\begin{bmatrix} 4 & 3 & 2 \\ -4 & 5 & 3 \end{bmatrix}$

41. $\begin{bmatrix} 20 \\ -24 \end{bmatrix}$

42. $\begin{bmatrix} 43 & -25 \\ 14 & 2 \end{bmatrix}$

43. $\begin{bmatrix} 7 & -3 & 4 \\ -1 & 2 & 3 \end{bmatrix}$

44. $\begin{bmatrix} 4 & 6 \\ 2 & 3 \end{bmatrix}$

45. 12

46. 27

47. 59

48. 170

POSTTEST ▶

If you have completed each of the 20 lessons in this book, then you are ready to complete the posttest in order to measure your progress on these concepts. The posttest, like the pretest, contains 50 multiple-choice questions based on the topics presented in this book. The structure is the same in both the pretest and the posttest; however, the questions are all different.

Take your time when completing the posttest. When you have it completed, check your answers with the key at the end of the posttest. In addition to the correct answer, the lesson number where the skills were taught is also listed, along with explanations for how the incorrect answers may have been attained. Use this information to assess your progress, and then compare your score on the posttest with your score on the pretest. If you scored higher on the posttest, great job! You have made progress while using this book, and should now focus on any questions you may have missed. Go back and review those topics, if any, in order to score even higher on your next placement exam.

If your score on the posttest was not as high as expected, take another look at the questions you missed. Did you miss a question because of a careless error? If so, then this type of error can be easily corrected by concentrating more on accuracy and reading the questions carefully. If you did not know how to complete the question you missed, go back to the lesson on that particular topic and spend more time working that type of problem. Take your time in understanding these important types of college placement test questions to help increase your achievement, and thus your score on the actual test! No matter what your score is on the posttest, keep this book at hand to use for future review and reference.

1. (a) (b) (c) (d)
2. (a) (b) (c) (d)
3. (a) (b) (c) (d)
4. (a) (b) (c) (d)
5. (a) (b) (c) (d)
6. (a) (b) (c) (d)
7. (a) (b) (c) (d)
8. (a) (b) (c) (d)
9. (a) (b) (c) (d)
10. (a) (b) (c) (d)
11. (a) (b) (c) (d)
12. (a) (b) (c) (d)
13. (a) (b) (c) (d)
14. (a) (b) (c) (d)
15. (a) (b) (c) (d)
16. (a) (b) (c) (d)
17. (a) (b) (c) (d)

18. (a) (b) (c) (d)
19. (a) (b) (c) (d)
20. (a) (b) (c) (d)
21. (a) (b) (c) (d)
22. (a) (b) (c) (d)
23. (a) (b) (c) (d)
24. (a) (b) (c) (d)
25. (a) (b) (c) (d)
26. (a) (b) (c) (d)
27. (a) (b) (c) (d)
28. (a) (b) (c) (d)
29. (a) (b) (c) (d)
30. (a) (b) (c) (d)
31. (a) (b) (c) (d)
32. (a) (b) (c) (d)
33. (a) (b) (c) (d)
34. (a) (b) (c) (d)

35. (a) (b) (c) (d)
36. (a) (b) (c) (d)
37. (a) (b) (c) (d)
38. (a) (b) (c) (d)
39. (a) (b) (c) (d)
40. (a) (b) (c) (d)
41. (a) (b) (c) (d)
42. (a) (b) (c) (d)
43. (a) (b) (c) (d)
44. (a) (b) (c) (d)
45. (a) (b) (c) (d)
46. (a) (b) (c) (d)
47. (a) (b) (c) (d)
48. (a) (b) (c) (d)
49. (a) (b) (c) (d)
50. (a) (b) (c) (d)

Posttest

1. What is the next prime number after 31?
  a. 33
  b. 35
  c. 37
  d. 41

2. Find the mean of the following set of numbers: 20, 17, 6, 13, 12, 12, 11, 19, and 7.
  a. 11
  b. 12
  c. 13
  d. 14

3. $0! = ?$
  a. 0
  b. 1
  c. 10
  d. 10.5

4. Find the mode for the following set of numbers: 20, 17, 6, 14, 14, 12, 11, 19, and 7.
  a. 6
  b. 12
  c. 13.3
  d. 14

5. Simplify $-|-8+3|$.
  a. 5
  b. −5
  c. 11
  d. −11

6. Simplify $(8 - 3) - (12 - 19 + 2)$.
  a. −24
  b. 0
  c. 10
  d. 14

7. Convert 0.0625 to a fraction.
  a. $\frac{1}{4}$
  b. $\frac{1}{8}$
  c. $\frac{1}{12}$
  d. $\frac{1}{16}$

8. What is the greatest common factor of 132 and 56?
  a. 4
  b. 6
  c. 8
  d. 22

9. Rebecca is traveling to Bangkok, Thailand. The exchange rate is 1 baht = 0.032 U.S. dollars. If Rebecca is bringing $1,300 on her trip, approximately how many baht will she receive at the exchange bureau?
  a. 40
  b. 35,000
  c. 40,000
  d. 45,000

10. Kevin bought Milky Way, Almond Joy, and Twix candy bars in a ratio of 7:4:3. There were 16 fewer Twix than Milky Way candy bars. What is the total number of candy bars purchased by Kevin?
  a. 4
  b. 14
  c. 40
  d. 56

11. Methane has the chemical formula CH_4. This means that each molecule of methane consists of one atom of carbon and four atoms of hydrogen. If a sample of methane contains 456 atoms of carbon, how many atoms of hydrogen does it contain?

a. 456

b. 1,824

c. 114

d. 2,280

12. Ms. Jenkins received a shipment of 3,600 juice glasses for her upcoming glassware sale. Inadequate packing material resulted in 12.5% breakage. How many juice glasses arrived damaged?

a. 225

b. 450

c. 720

d. 900

13. Naseem's school has 648 students, of whom 62.5% are girls. What fraction of the students are girls?

a. $\frac{3}{8}$

b. $\frac{2}{3}$

c. $\frac{3}{5}$

d. $\frac{5}{8}$

14. Order the following from the least to the greatest: $\frac{4}{15}$, 25%, $\frac{7}{16}$, 32%.

a. 25%, $\frac{4}{15}$, 32%, $\frac{7}{16}$

b. $\frac{7}{16}$, 25%, $\frac{4}{15}$, 32%

c. $\frac{7}{16}$, 32%, $\frac{4}{15}$, 25%

d. 25%, 32%, $\frac{4}{15}$, $\frac{7}{16}$

15. Simplify and write with only positive exponents: $(3x^{-4}y^4)^3$.

a. $\frac{3y^{12}}{x^{12}}$

b. $\frac{y^{12}}{x^{12}}$

c. $\frac{27y^{12}}{x^{12}}$

d. $\frac{27y}{x}$

16. Simplify by writing with no more than one radical: $\sqrt{12} \times \sqrt{30} \times \sqrt{18}$.

a. $1{,}296\sqrt{5}$

b. $4\sqrt{45}$

c. $18\sqrt{20}$

d. $36\sqrt{5}$

17. Which of the following is equal to 1.234×10^{-2}?

a. 0.01234

b. 0.001234

c. 123.4

d. 123,400

18. Simplify the expression: $0.6x^2 + 1.3x + 0.15x^2 - 0.2x$.

a. $0.75x^2 + 1.5x$

b. $0.45x^2 + 1.45x$

c. $0.75x^2 + 1.1x$

d. $0.8x^2 + 1.45x$

19. Perform the indicated operations and combine like terms: $(7a^2 - b) - (5a^2 + 2b)$.

a. $2a^2 + b$

b. $2a^2 + 3b$

c. $2a^2 - b$

d. $2a^2 - 3b$

20. Evaluate $4(x^2 + 3x) - 5y$, when $x = 2$ and $y = -7$.

a. 5

b. 179

c. 75

d. 49

21. Alessandra is making and selling friendship bracelets. She has calculated that the colored string and beads per bracelet cost her approximately $0.45. How many bracelets will Alessandra have to sell for $1.00 each to make a profit of $25.30?

    a. 26

    b. 46

    c. 56

    d. 14

22. Solve the following equation for x: $4 - 5x = \frac{3x}{5}$.

    a. $\frac{1}{2}$

    b. 2

    c. $\frac{5}{7}$

    d. $\frac{7}{5}$

23. Solve the following equation for x:
$5x = 4(2x + 6)$.

    a. 8

    b. −8

    c. −2

    d. 2

24. Which of the following is a solution for the inequality $7x - 18 > 31$?

    a. 8

    b. 7

    c. 0

    d. −3

25. Solve $5x \leq -\frac{1}{2}$.

    a. $x \leq 10$

    b. $x \leq \frac{1}{10}$

    c. $x \leq -\frac{1}{10}$

    d. $x \geq -\frac{1}{10}$

26. If angle A in an isosceles triangle is 24° greater than angle B, and angle C is congruent to angle B, what are the measurements of all three angles of the triangle?

    a. 38°, 38°, and 104°

    b. 52°, 52°, and 76°

    c. 72°, 68°, and 40°

    d. 44°, 44°, and 68°

27. In the following figure, the triangles are similar. Find x.

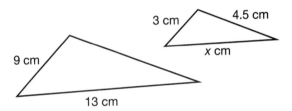

    a. $\frac{3}{13}$ cm

    b. $\frac{1}{3}$ cm

    c. 39 cm

    d. $\frac{13}{3}$ cm

28. Find the value of the angle measures x, y, and z.

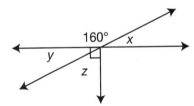

    a. 40°, 40°, 50°

    b. 20°, 20°, 80°

    c. 20°, 20°, 70°

    d. 20°, 45°, 45°

29. If a circle has an area of 1,256 mm², what is its diameter?

    a. 40 mm

    b. 35 mm

    c. 20 mm

    d. 10 mm

30. Find the length of the hypotenuse to the nearest hundredth in a right triangle with leg lengths of 7 units and 13 units.

a. 10.95

b. 14.22

c. 17.78

d. 14.76

31. If a triangle has sides that are 13 cm, 15 cm, and 17 cm long, what is the perimeter of the triangle?

a. 38 cm

b. 45 cm

c. 78 cm

d. 90 cm

32. What is the distance, to the nearest hundredth, between the points (–2,5) and (4,–10) on the coordinate plane?

a. 16.16 units

b. 15.13 units

c. 5.39 units

d. 7.81 units

33. What is the equation of the line graphed on this coordinate system?

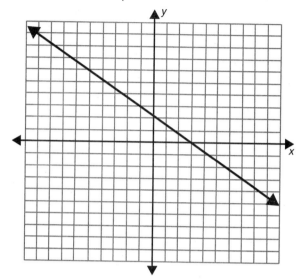

a. $y = \frac{3}{4}x + 2$

b. $y = -\frac{3}{4}x + 2$

c. $y = -3x + 4$

d. $y = -\frac{4}{3}x + 2$

34. Find the solution to the system of equations in this figure.

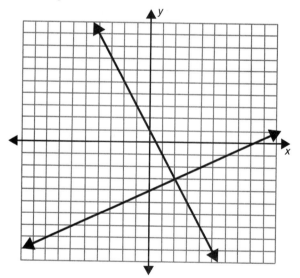

a. (2,–3)

b. (–3,2)

c. (1,–4)

d. (8,0)

35. Find the solution to the system of equations $y = x + 3$ and $2y + x = 12$.

    a. (5,2)

    b. (4.5,7.5)

    c. (2,5)

    d. (3,6)

36. Factor $28x^2 + 49x + 21$ completely.

    a. $7(2x + 3)(2x + 1)$

    b. $7(4x + 1)(x + 3)$

    c. $7(4x^2 + 7x + 3)$

    d. $7(4x + 3)(x + 1)$

37. Expand (multiply and simplify): $(2x + 3)^2$.

    a. $4x + 6$

    b. $4x^2 + 9$

    c. $4x^2 + 6x + 9$

    d. $4x^2 + 12x + 9$

38. Which of the following describes the end behavior of $f(x) = 5x^7 + 3x^5 - x^3 + 1{,}333$?

    a. The ends go in opposite directions. The graph enters at the top and exits at the bottom.

    b. The ends go in the same direction. Both point upward.

    c. The ends go in the same direction. Both point downward.

    d. The ends go in opposite directions. The graph enters at the bottom and exits at the top.

39. Simplify the following rational expression to lowest terms: $\frac{x^2 - 3x - 40}{x^2 - 10x + 16}$.

    a. $\frac{(x + 5)(x + 8)}{(x - 8)(x - 2)}$

    b. $\frac{(x + 8)}{(x - 8)}$

    c. $\frac{(x - 5)(x + 8)}{(x - 8)(x - 2)}$

    d. $\frac{(x + 5)}{(x - 2)}$

40. Solve for all values of x: $x^2 + 4x = 12$.

    a. $\{-2, -6\}$

    b. $\{2, -6\}$

    c. $\{-2, 6\}$

    d. $\{2, 6\}$

41. Given $f(x) = -3x^2 + 8x - 4$ and $g(x) = 6x - 8$, find $g[f(3)]$.

    a. -7

    b. -50

    c. -224

    d. 47

42. Find the inverse of the function $y = -\frac{3}{4}x + 3$.

    a. $y = -\frac{4}{3}x - 3$

    b. $y = x - 3$

    c. $y = -\frac{4}{3}x + 4$

    d. $y = -4x - 12$

43. Solve for n: $\log_5 n - \log_5 \frac{1}{3} = \log_5 30$.

    a. $n = 90$

    b. $n = 29\frac{2}{3}$

    c. $n = 30\frac{1}{3}$

    d. $n = 10$

44. Solve for x. Round to the nearest hundredth if necessary: $12^x = 36$.

    a. $x \approx 1.44$

    b. $x = 3$

    c. $x = 24$

    d. $x = 1.56$

45. If $\cos \theta = \frac{8}{17}$ and $\tan \theta < 0$, then $\sin \theta = ?$

    a. $\frac{17}{15}$

    b. $-\frac{15}{17}$

    c. $-\frac{17}{15}$

    d. $\frac{15}{17}$

46. What is the amplitude of the graph of $f(x) = 4 \sin 2x$?

    a. 2

    b. π

    c. 2π

    d. 4

47. $\csc \theta = ?$

    a. $\frac{1}{\sec \theta}$

    b. $\frac{1}{\cos \theta}$

    c. $\frac{1}{\sin \theta}$

    d. $\frac{\cos \theta}{\sin \theta}$

48. Simplify the following trigonometric expression to a single function: $\tan x \csc x$.

    a. $\tan x$

    b. $\csc x$

    c. $\sec x$

    d. $\cot x$

49. For $f(x) = \sin\frac{1}{4}x$, what does the $\frac{1}{4}$ indicate?

    a. The amplitude of the function is $\frac{1}{4}$.

    b. The period is $\frac{1}{4}\pi$.

    c. The frequency is $\frac{1}{4}\pi$.

    d. The frequency is $\frac{1}{4}$.

50. Find the 12th element in the geometric sequence $A_n = \{3, 6, 12, 24, 48, \ldots\}$.

    a. 93

    b. 6,144

    c. 3

    d. 12,285

Answers

1. c. (Lesson 1) Prime numbers are divisible only by 1 and themselves. Answer choices **a** and **b** are therefore incorrect because 33 is divisible by 3 and 11, and 35 is divisible by 5 and 7. Answer choices **c** and **d** are each prime numbers. Answer choice **c** is the next prime after 31.

2. c. (Lesson 1) The mean of this set of numbers is 13. The mean is an average of a set of numbers. Add up the numbers and divide by the total number of numbers in the set. There are nine numbers in this set.
$20 + 17 + 6 + 13 + 12 + 12 + 11 + 19 + 7 = 117$
$117 \div 9 = 13$
Choice **a** is incorrect because the mean of this set of numbers is 13, not 11, and could be the result of a calculation error. Choice **b** is incorrect because the mean of this set of numbers is 13, not 12, which is the mode (and the median) of the set. Choice **d** is the range (highest value – lowest value) of the set.

3. b. (Lesson 1) There are several proofs that show $0! = 1$. Think of it this way: a factorial calculates the number of ways a set of n numbers can be combined (how many permutations there are). Even an empty set can be ordered in one way; thus $0! = 1$. The other choices are incorrect. Choice **a** is incorrect because every set can be combined in at least one way. Choices **c** and **d** are incorrect because the set is empty, so it can be combined in only one way.

4. d. (Lesson 1) The mode for this set of numbers is 14. The mode is the value that appears most often in the set of numbers. If a set of numbers has no repeated values, there is no mode. The number 14 appears most often and is therefore the mode for this set. Choice **a** is incorrect and is the lowest value in the set. Choice **b** is incorrect and appears only once in the set. Choice **c** is incorrect because 13.3 is the mean of the set, not the mode.

5. b. (Lesson 2) You perform the calculation inside the absolute value sign first: $-8 + 3 = -5$, and the absolute value of $-5 = 5$. Therefore, the answer is -5. Choice **a** is incorrect because it neglects the negative sign on the outside of the absolute value. Choices **c** and **d** are incorrect because $-8 + 3 = -5$, not 11 or -11.

6. c. (Lesson 2) First, evaluate what is in parentheses: $8 - 3 = 5$; then evaluate the second set of parentheses: $12 - 19 = -7$, and $-7 + 2 = -5$. Now, perform the subtraction to get $5 - (-5) = 10$, choice **c**. If your choice was **a**, you ignored the parentheses and just went left to right. If your choice was **b**, you incorrectly evaluated $5 - (-5)$ as 0. If you chose **d**, you incorrectly performed addition before subtraction in the second set of parentheses.

7. d. (Lesson 3) $0.0625 = \frac{625}{10,000}$. Dividing by $\frac{25}{25}$ to simplify the fraction, $\frac{625}{10,000} = \frac{25}{400}$. Dividing by $\frac{25}{25}$ again, $\frac{25}{400} = \frac{1}{16}$. Or simply divide by $\frac{625}{625}$ to yield $\frac{1}{16}$, but it would be unusual to know that 625 is a factor of 10,000. Another way to find the correct answer is to divide the numerator by the denominator in each of the fraction choices: $1 \div 16 = 0.0625$. Choice **a** is incorrect: $\frac{1}{4} = 0.25$, not 0.0625. Choice **b** is incorrect: $\frac{1}{8} = 0.125$, not 0.0625. Choice **c** is incorrect: $\frac{1}{12} = 0.0833\ldots$

8. a. (Lesson 3) The factors of 132 are: $\{1, 2, 3, 4, 6, 11, 12, 22, 33, 44, 66, 132\}$. The factors of 56 are: $\{1, 2, 4, 7, 8, 14, 28, 56\}$. The greatest common factor (GCF) of the two numbers is 4. Alternatively, you can look at the prime factorization of each number and multiply the common prime factors. The prime factors of 56 are $2 \times 2 \times 2 \times 7$. The prime factors of 132 are $2 \times 2 \times 3 \times 11$. Multiplying the common prime factors (2×2), the GCF is 4. Choice **b** is incorrect: 6 is not a factor of 56. Choice **c** is incorrect: 8 is not a factor of 132. Choice **d** is incorrect: 22 is not a factor of 56.

9. c. (Lesson 4) For approximately every 3 cents, Rebecca receives 1 baht. 13,000 cents therefore yields her 39,000 bahts, closest to choice **c**, 40,000. To confirm, set up the following proportion: $\frac{1}{0.032} = \frac{x}{1,300}$, where x is the unknown number of bahts. Cross multiplying yields $0.032x = 1,300$. Then, $x = 1,300 \div 0.032 = 40,625$, closest to answer choice **c**, 40,000. If your choice was **a**, you set up the proportion incorrectly. Choice **b** is incorrect because Rebecca will receive approximately 40,000 baht at the exchange bureau, not 35,000. If your choice was **d**, that is incorrect because you may have rounded 40,625 incorrectly to 45,000.

10. d. (Lesson 4) The candy bars are in the ratio of $7x$ to $4x$ to $3x$. The difference in the number of Milky Ways purchased and the number of Twix candy bars purchased is 16. Set up the equation $7x = 3x + 16$. Subtract $3x$ from both sides to get $4x = 16$, and $x = 4$. Multiply each part of the ratio by 4 to get 28:16:12. The total number of candy bars purchased was answer choice **d**, 56. Choice **a** is incorrect because it is the value of x, the multiplier, not the total number of candy bars. Choice **b** is incorrect because it is not the total number of candy bars; it is the sum of 7, 4, and 3. If your choice was **c**, you gave the amount of Twixes and Milky Ways, not the total number of candy bars.

11. b. (Lesson 4) Set up the proportion $\frac{1}{4} = \frac{456}{x}$, where x is the unknown number of hydrogen atoms. Cross multiplying yields $x = 456 \times 4 = 1,824$. Choice **a** is the number of carbon atoms, not the hydrogen atoms. If your choice was **c**, you set up the proportion incorrectly. If you chose answer choice **d**, you gave the total number of atoms, both carbon and hydrogen, not just the hydrogen atoms.

12. b. (Lesson 5) To determine the number of glasses damaged, multiply 3,600 by 12.5%: $3,600 \times 0.125 = 450$. Choice **a** is incorrect. You can quickly eliminate answer choices by realizing that 10% of $3,600 = 360$ (allowing you to eliminate choice **a**, since 225 is less than this number). Choice **c** is incorrect. Since 10% of $3,600 = 360$, then 20% of 3,600 is 720. This also allows you to eliminate choice **d**, which is greater than this number.

13. d. (Lesson 5) Because 62.5% of the students are girls, you can express this percentage as a fraction and simplify:
$\frac{62.5}{100} = \frac{625}{1,000} \div \frac{125}{125} = \frac{5}{8}$
Choice **a** is incorrect: If 62.5% of the students are girls, then more than half are girls; $\frac{3}{8}$ is less than half, so you can eliminate it. Choice **b** is incorrect: $\frac{2}{3} = 0.66\ldots$, which is more than 0.625 (or 62.5%). Choice **c** is incorrect: $\frac{3}{5} = 0.60$, or 60%, which is less than 0.625 (or 62.5%).

14. a. (Lesson 5) To determine the answer, convert the fractions to percentages so you can easily compare: $\frac{4}{15} = 26.67\%$; $\frac{7}{16} = 43.75\%$. The correct order is 25%, 26.67%, 32%, 43.75%, or 25%, $\frac{4}{15}$, 32%, $\frac{7}{16}$. Choices **b** and **d** are incorrect because they are in the incorrect order. Choice **c** is in order from greatest to least.

15. c. (Lesson 6) $(3x^{-4}y^4)^3 = 3^3x^{-12}y^{12} = 27(\frac{1}{x^{12}})y^{12} = \frac{27y^{12}}{x^{12}}$. Choice **a** is incorrect because it neglects to raise 3 to the third power in the first step of the simplification: $(3x^{-4}y^4)^3 = 3^3x^{-12}y^{12}$. Choice **b** is incorrect because it leaves out the coefficient of $3^3 = 27$. If you chose answer choice **d**, you thought that you could cancel out the exponents. You can subtract exponents only when the base variables are the same.

16. d. (Lesson 6) First, write each term with prime factorization:

$$\sqrt{12} = \sqrt{2 \times 2 \times 3}$$
$$\sqrt{30} = \sqrt{2 \times 3 \times 5}$$
$$\sqrt{18} = \sqrt{2 \times 3 \times 3}$$

Put them together and pair up similar primes:

$$\sqrt{12 \times 30 \times 18} =$$
$$\sqrt{2 \times 2 \times 2 \times 2 \times 3 \times 3 \times 3 \times 3 \times 5}$$

Two pairs of 2's and two pairs of 3's can come out of the radical sign to get $2 \times 2 \times 3 \times 3\sqrt{5} = 36\sqrt{5}$.

If your answer was choice **a**, you did not take out the prime pairs correctly; you took out all of the primes that had pairs, not just one prime for each pair. If you chose answer **b**, you did not take out the second pair of 3's. If your answer was choice **c**, you did not take out the second pair of 2's.

17. a. (Lesson 6) When there is a negative power of 10, you move the decimal point to the left. In this case, you move it left two places to get 0.01234. If your choice was **b**, you incorrectly placed two zeros to the front of the significant digits, instead of moving the decimal two places. If you chose answer **c**, you moved the decimal point two places to the right, instead of two digits to the left. If your answer choice was **d**, you incorrectly moved the decimal point to the right instead of left, and also incorrectly thought that the exponent 2 signifies two zeros.

18. c. (Lesson 7) Combine like terms: $0.6x^2 + 0.15x^2 = 0.75x^2$. Then, $1.3x - 0.2x = 1.1x$. The simplified expression is $0.75x^2 + 1.1x$, which is answer choice **c**. Choice **a** is incorrect—it improperly combines $1.3x$ and $-0.2x$. Choices **b** and **d** are also incorrect and are also the result of incorrectly combining terms.

19. d. (Lesson 7) When performing subtraction of polynomials, add the opposite of the second polynomial. Combine like terms: $7a^2 - 5a^2 = 2a^2$; $-b - 2b = -3b$, leaving answer choice **d**, $2a^2 - 3b$. Choice **a** is incorrect because it neglects to carry the negative sign over to the term $2b$. Choice **b** is incorrect because $-b - 2b = -3b$, not $3b$. Choice **c** is incorrect because $-b - 2b = -3b$, not $-b$.

20. c. (Lesson 7) Substitute the values of x and y into the equation and simplify using the order of operations rules: $4(4 + 6) - 5(-7) = 4(10) - (-35) = 40 + 35 = 75$, which is answer choice **c**. Choice **a** is incorrect; 35 was subtracted from 40 instead of added. Choice **b** is incorrect; if you arrived at this answer, you evaluated $3x$ as 32 instead of 3×2. If your answer was choice **d**, you added 4 to 10 instead of multiplying.

21. b. (Lesson 8) If Alessandra sells each bracelet for $1.00 and the cost of materials is $0.45, she makes a profit of $0.55 per bracelet: $25.30 \div 0.55 = 46$. Choice **a** is incorrect—if she sold 26 bracelets she would earn only $14.30 after her expenses. If your answer was choice **c**, you incorrectly calculated using $0.45 as her profit, instead of $0.55. Choice **d** is incorrect. If you chose this answer, you multiplied by $0.55 instead of dividing.

22. c. (Lesson 8) First, multiply both sides by 5 to clear the fraction:

$$5(4 - 5x) = \frac{5 \times 3x}{5}$$

Use the distributive property to simplify: $20 - 25x = 3x$. Add $25x$ to both sides to get $20 = 28x$. Divide both sides by 28 to get $\frac{20}{28} = x$. Simplify the fraction to get $\frac{5}{7}$. If you chose answer choice **a**, you disregarded the 5 in the denominator. If your answer was **b**, not only did you disregard the 5 in the denominator, but you also divided in the incorrect order. If you chose **d**, you divided in the incorrect order.

23. b. (Lesson 8) The first step to solve $5x = 4(2x + 6)$ is to use the distributive property: $5x = 8x + 24$. Subtract $8x$ from both sides of the equation to get the variable on one side, and $-3x = 24$. Divide both sides by -3 to get $x = -8$. If your answer was choice **a**, you incorrectly did integer division. If you chose choice **c**, you did not distribute the 4 to the second term in parentheses. If your answer was choice **d**, you ignored the factor of 4 altogether.

24. a. (Lesson 9) Solve the inequality by isolating the variable:

$7x > 31 + 18$

$7x > 49$

$x > \frac{49}{7}$

$x > 7$

The only answer greater than 7 is answer choice **a**, 8. Choice **b** is incorrect because it is equal to 7. Choices **c** and **d** are incorrect because they are each less than 7.

25. c. (Lesson 9) You solve linear inequalities the same way as linear equations, with several exceptions. The answer in this format reflects that it is the set of all x values such that x is a real number that is less than, less than or equal to, greater than, or greater than or equal to a real number. So for $5x \leq \frac{1}{2}$, divide both sides of the inequality by 5 to get $x \leq -\frac{1}{2} \div \frac{5}{1} = x \leq -\frac{1}{2} \times \frac{1}{5} = x \leq -\frac{1}{10}$. Or alternatively, multiply both sides of the equation by 2 to get $10x \leq -1$, which is equivalent to $x \leq -\frac{1}{10}$. Choice **a** is incorrect and is the result of multiplying -2 by 5 to isolate the variable. Choice **b** is incorrect. It neglects to keep the negative value on the right side of the inequality. Choice **d** is incorrect. When solving inequalities, if you multiply or divide both sides by a negative in order to get a positive variable, you must flip the inequality sign. In this problem, you either divided by $+5$ or multiplied by $+2$, so the inequality should not reverse.

26. b. (Lesson 10) Isosceles triangles have two equal angles. Since you know that the sum of the interior angles in a triangle is 180°, look at the answer choices and eliminate any combinations that do not add to 180°. Choice **d** is incorrect because the sum of the angles is 156°, not 180°. Choice **c** is not correct, because these angle measures cannot be an isosceles triangle, as none of the angles have the same measure. Choice **a** is incorrect. Even though the angles add to 180° and there are two angles with the same measure, they do not differ by 24: $104 - 38 = 66$, not 24.

27. d. (Lesson 4) Remember the concept that similar triangles have proportional sides. Simply set up the proportion and solve:

$$\frac{long_{bigT}}{short_{bigT}} = \frac{long_{small\,t}}{short_{small\,t}}$$

$$\frac{13}{9} = \frac{x}{3}$$

$$9x = 39$$

$$x = \frac{39}{9} = \frac{13}{3}$$

Choice **a** is incorrect—the fraction is flipped, so you probably set up the proportion incorrectly. If your answer was choice **b**, you used the side on the small triangle measuring 4.5 instead of using the measure of 3. If your answer was choice **c**, you multiplied 13 by 3 instead of dividing by 3.

28. c. (Lesson 10) The angle marked as x is 20°, because it is a linear pair with the angle marked as 160°, and 180° − 160° = 20°. The angle marked y is also 20°, because it is a vertical angle to angle x, and vertical angles are congruent. The angle marked z is 70°, because it is complementary to the angle marked as y, 20°. 90° − 20° = 70°. If you chose choice **a**, you may have incorrectly thought that a linear pair of angles sums to 200, instead of 180. If your answer was choice **b**, you had the angles x and y correct, but you may have incorrectly thought that a complementary pair of angles sums to 100, instead of 90. If you chose answer choice **d**, you had angle x correct, but you may have incorrectly thought that the two angles forming the right angle were congruent.

29. a. (Lesson 11) The formula for the area of a circle is $A = (\pi)(r^2)$. You know that $\pi \approx 3.14$ and $d = 2r$. You can solve as follows:

$1{,}256 = (3.14)(r^2)$

$\frac{1{,}256}{3.14} = r^2$

$400 = r^2$

$\sqrt{400} = r$

$20 = r$

$40 = d$

Therefore, answer choice **a**, 40 mm, is the correct answer. Choice **b** could be the result of a calculation error and choice **c** is incorrect because 20 is the radius, not the diameter. Selecting choice **d** could be the result of taking half of the diameter twice.

30. d. (Lesson 11) To calculate the length of the hypotenuse of a right triangle, you use the Pythagorean theorem, which states that the square of the hypotenuse is equal to the sum of the squares of the other two sides of the triangle. The formula is $a^2 + b^2 = c^2$, where c is the longest side of the right triangle and a and b are the other two sides. For this problem:

$7^2 + 13^2 = c^2$

$49 + 169 = 218$

$\sqrt{218} = \sqrt{c^2}$

$14.76482306 = c$

Therefore, answer choice **d**, 14.76, is the correct answer. Selecting answer choice **a** could be the result of using 13 as the hypotenuse of the right triangle. Choices **b** and **c** are also incorrect and could be the result of errors in calculation.

31. b. (Lesson 11) To calculate the perimeter, add up the lengths of each side of a figure. The perimeter of this triangle is thus 13 + 15 + 17 = 45. Choices **a** and **c** are incorrect and could be the result of an error in calculation. Choice **d** is the result of multiplying each side by 2 before adding the sides together.

32. a. (Lesson 12) Use the distance formula: $d = \sqrt{(x_2 - x_1)^2 + (y_2 - y_1)^2}$, where $(x_1, y_1) = (-2, 5)$ and $(x_2, y_2) = (4, -10)$. By substitution, $d = \sqrt{(4 - -2)^2 + (-10 - 5)^2} = \sqrt{6^2 + (-15)^2} = \sqrt{36 + 225} = \sqrt{261} \approx 16.16$. If you chose **b**, you incorrectly subtracted 4 and −2 for the wrong result of 2 instead of 6. If your answer was choice **c**, you incorrectly subtracted both 4 and −2 for the wrong result of 2 instead of 6, and −10 and 5 for the wrong result of 5 instead of −15. If your answer was **d**, you incorrectly subtracted −10 and 5 for the wrong result of 5 instead of −15.

33. b. (Lesson 12) In the figure, the y-intercept is +2. The slope is determined by $\frac{rise}{run}$. On the graph, to get from one point to the next point that has integer coordinates, you go down three units and four units to the right, or alternatively, up three units and four units to the left. This is a rise of 3 and a run of 4, with the slope being negative. The slope is therefore $-\frac{3}{4}$. So the equation is $y = -\frac{3}{4}x + 2$. If your answer was choice **a**, you did not recognize that the slope was negative. If you chose answer choice **c**, you used the rise and run parts of the slope ratio, which should have made up the coefficient of x, as the both the coefficient and the y-intercept portions of the equation. If the answer was **d**, you exchanged the rise with the run, to get the wrong slope in the equation.

34. a. (Lesson 13). To find the solution to the system of equations graphically, you name the coordinates of the intersection of the two graphed lines. This intersection point is $(2,-3)$. If your answer was **b**, you flipped the x and y coordinates. If your answer was choice **c**, you named the y-intercepts of the two graphed lines, instead of the intersection point. Answer choice **d** is incorrect; this is the x-intercept of one of the graphed lines.

35. c. (Lesson 13) The find the solution to the system of equations $y = x + 3$ and $2y + x = 12$, one method is to substitute in $(x + 3)$ for y in the second equation.

$2y + x = 12$ Write the original second equation.

$2(x + 3) + x = 12$ Substitute in for y.

$2x + 6 + x = 12$ Distribute the 2.

$3x + 6 = 12$ Combine like terms.

$-6 = -6$ Subtract 6 from both sides.

$3x = 6$

$\frac{3x}{3} = \frac{6}{3}$ Divide both sides by 3.

$x = 2$

Substitute in the value of 2 for x in the first equation, and $y = 2 + 3 = 5$. The ordered pair is $(2,5)$. If your answer was **a**, you probably solved the system correctly, but switched the x and y coordinate in the answer. If your answer was choice **b**, you may have written the wrong substitution equation as $2x + 3 = 12$ and solved to get $x = 4.5$. If you chose answer choice **d**, you forgot to distribute the 2 to the 3, or you may have forgotten to combine like terms, getting $2x$ instead of the correct term, $3x$.

36. d. (Lesson 14) When factoring polynomials, the first step is always to find the greatest common factor (GCF) before you try to do anything else. In this problem, 7 is the GCF: $28x^2 + 49x + 21 = 7(4x^2 + 7x + 3)$. Next, try to find binomial factors. This may take some trial and error. The trinomial $(4x^2 + 7x + 3)$ can be factored as follows: $(4x^2 + 7x + 3) = (4x + 3)(x + 1)$. The solution is: $28x^2 + 49x + 21 = 7(4x^2 + 7x + 3) = 7(4x + 3)(x + 1)$. Choices **a** and **b** are incorrect because the trinomials are incorrectly factored after dividing by 7. Choice **c** is incorrect because it is not completely factored.

37. d. (Lesson 14) Simplify as follows: $(2x + 3)^2 = (2x + 3)(2x + 3)$. Now, use the distributive property (FOIL) to multiply the binomials. $(2x + 3)(2x + 3) = 4x^2 + 6x + 6x + 9$. Combine like terms to get $4x^2 + 12x + 9$. Choice **a** is incorrect because it simply doubles each term inside the parentheses. Choice **b** is incorrect because it only multiplies the first terms in parentheses and the last terms in parentheses, and does not multiply the inner and outer terms together. Choice **c** is incorrect because it incorrectly combines the like terms of $6x$ and $6x$.

38. d. (Lesson 14) The leading term, $5x^7$, contains an odd-degree exponent; therefore, the ends go in opposite directions. That makes choices **b** and **c** incorrect. The coefficient is positive, so it enters the graph from the bottom and exits the graph at the top. That makes choice **a** incorrect.

39. d. (Lesson 15) Simplify as follows by factoring and canceling out any common factors from the numerator and denominator:
$$\frac{x^2 - 3x - 40}{x^2 - 10x + 16} = \frac{(x + 5)(x - 8)}{(x - 8)(x - 2)} = \frac{(x + 5)}{(x - 2)}$$
Choice **a** is incorrect. It improperly factors $x^2 - 3x - 40$ to $(x + 5)(x + 8)$, not $(x + 5)(x - 8)$. Choice **b** is incorrect. It cancels out the terms that are *not* alike ($x + 5$ and $x - 2$), not the terms that are alike: $x - 8$, $x - 8$. Choice **c** is incorrect. It improperly factors $x^2 - 3x - 40$ to $(x - 5)(x + 8)$, not $(x + 5)(x - 8)$.

40. b. (Lesson 15) First, get the quadratic equation into standard form: $x^2 + 4x = 12$ becomes $x^2 + 4x - 12 = 0$. Next, factor the left side of the equation to get $(x + 6)(x - 2) = 0$. Set each factor equal to zero and solve for x. $x + 6 = 0$, so $x = -6$; $x - 2 = 0$, so $x = 2$. Answer choices **a**, **c**, and **d** are all errors in the signs of the solutions.

41. b. (Lesson 16) First you evaluate function f when $x = 3$: $-3(3)^2 + 8(3) - 4 = -3 \times 9 + 24 - 4 = -27 + 24 - 4 = -3 - 4 = -7$. Take this result, -7, and evaluate function g when $x = -7$: $(6 \times -7) - 8 = -42 - 8 = -50$. If your answer was **a**, you only evaluated $f(3)$ and did not take that result and evaluate $g(x)$. If you chose answer choice **c**, you evaluated the functions in the opposite order; you must evaluate f first and then g. If your answer was **d**, you incorrectly evaluated $-3(3)^2$ as $+27$ when the correct value is -27.

42. c. (Lesson 16) To find the inverse of a function, exchange the x and y variables, and solve for y.

$y = -\frac{3}{4}x + 3$	Write the original equation.
$x = -\frac{3}{4}y + 3$	Exchange the x and y variables.
$\dfrac{-3 = -3}{x - 3 = -\frac{3}{4}y}$	Subtract 3 from both sides. Simplify.
$-\frac{4}{3}(x - 3) = -\frac{4}{3}(-\frac{3}{4}y)$	Multiply both sides by $-\frac{4}{3}$.
$y = -\frac{4}{3}x + 4$	Simplify using the distributive property.

If your answer was **a**, you did not distribute the $-\frac{4}{3}$ to the negative 3. If you chose answer choice **b**, you ignored the coefficient of $-\frac{3}{4}$. If you chose **d**, you ignored the denominator of 3 when undoing the coefficient.

43. d. (Lesson 17) Use the Laws of Logarithms.

$\log_5 n - \log_5 \frac{1}{3} = \log_5 30$ Write the original equation.

$\log_5\left(\frac{n}{\frac{1}{3}}\right) = \log_5 30$ Use the Laws of Logarithms for division.

$3n = 30$ Undo the log from both sides and simplify the complex fraction.

$n = 10$ Divide both sides by 3.

If your answer was choice **a**, you multiplied by 3 instead of dividing by 3. You many have simplified the complex fraction incorrectly. If you chose choice **b**, you incorrectly subtracted one-third from 30. If your answer was choice c, you incorrectly added one-third to 30.

44. a. (Lesson 17) To solve this equation, take the common logarithm of both sides:

$12^x = 36$ Write the original equation.

$\log 12^x = \log 36$ Take the common logarithm of both sides.

$x\log 12 = \log 36$ Use the Exponent Rule for Logarithms.

$x = \frac{\log 36}{\log 12}$ Divide both sides by the common logarithm of 12.

$x \approx 1.44$ Divide and simplify.

If your answer was **b**, you divided both sides by 12, thinking the equation was $12x = 36$. If you chose answer choice **c**, you subtracted 12 from both sides instead of taking the common logarithm of both sides. Choice **d** is incorrect; this is the common logarithm of 36 and you forgot to divide this by the common logarithm of 12.

45. b. (Lesson 19) When you are given a problem that indicates the sign for one of the trigonometric functions, it means that you are working within the four quadrants of the rectangular coordinate system. There are certain facts you need to know in order to get these problems correct. Sine corresponds to the y-axis and cosine corresponds to the x-axis. Thus, if x is positive, cosine is positive, but if x is negative, cosine is negative. On the other hand, if y is positive, sine is positive, but if y is negative, sine is negative. Since $\tan = \frac{\sin}{\cos}$, when sin and cos are the same sign, tan is positive. When sin and cos are opposite signs, tan is negative. If a function is positive, its reciprocal is positive. If a function is negative, its reciprocal is negative. Thus, in the second quadrant, only sine and its reciprocal, cosecant, are positive. In the third quadrant, only tangent and its reciprocal, cotangent, are positive. Finally, in the fourth quadrant, only cosine and its reciprocal, secant, are positive. This is probably your best memory aid. QI: all positive, QII: only sine positive, QIII: only tan positive, QIV: only cosine positive. Just remember that all other functions are negative in that quadrant and that the reciprocals have the same sign.

In this problem, you know that $\tan \theta < 0$ and $\cos \theta = \frac{8}{17}$. Since QIV: only cosine positive, you know that the $\sin \theta < 0$. The terminal side of the angle, or the ray where the measurement of the angle stops, is in quadrant IV. Since $\cos \theta = \frac{adjacent}{hypotenuse} = \frac{8}{17}$, you know two sides of the triangle. Use the Pythagorean theorem to find the third.

$a^2 + 8^2 = 17^2$

$a^2 + 289 - 64$

$a^2 = 225$

$a^2 = \pm 15$

Since $\sin\theta = \frac{opposite}{hypotenuse}$, $\sin = \frac{\pm 15}{17}$. Since it is in the fourth quadrant, the sine is negative. The correct answer is choice **b**: $-\frac{15}{17}$.

Choice **a** is incorrect because the angle is in the fourth quadrant, not the first. Choice **c** is incorrect because the operation is sine, not cosecant. Choice **d** is incorrect because it has the incorrect sign on the value.

46. d. (Lesson 18) The normal range for both sine and cosine is $-1 \leq y \leq 1$. When a coefficient is placed in front of the function, it stretches the graph vertically. In other words, it increases the amplitude. Thus, the amplitude of $y = 4\sin 2x$ is 4 and its range is $-4 \leq y \leq 4$. Amplitude is always a positive quantity (amplitude $= |a|$). Choice **a** is the frequency of the graph. Choice **b** is the distance between any two repeating points on this function that is a sine function with a frequency of 2. Choice **c** is the distance between any two repeating points on a sine function with a frequency of 1.

47. c. (Lesson 19) This problem requires that you know the common basic (reciprocal) trigonometric identities. They are as follows:

$\sin\theta = \frac{1}{\csc\theta}$ $\csc\theta = \frac{1}{\sin\theta}$(cosecant) $= \frac{hypotenuse}{opposite}$

$\cos\theta = \frac{1}{\sec\theta}$ $\sec\theta = \frac{1}{\cos\theta}$(secant) $= \frac{hypotenuse}{adjacent}$

$\tan\theta = \frac{1}{\cot\theta} = \frac{\sin\theta}{\cos\theta}$ $\cot\theta = \frac{1}{\tan\theta} = \frac{\cos\theta}{\sin\theta}$(cotangent) $= \frac{adjacent}{opposite}$

Choices **a**, **b**, and **d** are all incorrect because they are not the correct reciprocal trigonometric identity for cosecant θ.

48. c. (Lesson 19) To simplify the expression $\tan x \csc x$, use one of the identities, which you must memorize. This problem also requires using one of the common basic (reciprocal) trigonometric identities, which you should also have memorized. These are as follows:

$\sin x = \frac{1}{\csc x}$ $\csc x = \frac{1}{\sin x}$

$\cos x = \frac{1}{\sec x}$ $\sec x = \frac{1}{\cos x}$

$\tan x = \frac{1}{\cot x} = \frac{\sin x}{\cos x}$ $\cot x = \frac{1}{\tan x} = \frac{\cos x}{\sin x}$

Cotangent is the reciprocal of tangent, cosecant is the reciprocal of sine, and cosine is the reciprocal of secant. Substitute using the identities: $\tan x \csc x = \frac{\sin x}{\cos x} \times \frac{1}{\sin x}$. Cancel the common factor of $\sin x$ to get $\frac{1}{\cos x} = \sec x$. Choice **a** is incorrect because it is only the $\tan x$. Choice **b** is incorrect because it names cosecant, which is the wrong function. Choice **d** is incorrect because the cotangent function is not used in simplifying this expression.

49. d. (Lesson 18) In this function, **a** represents the amplitude of the function, which is the highest and lowest the function will go. The value of **b** in the equation represents the frequency. In this function, the amplitude is 1 and the frequency is $\frac{1}{4}$, so choice **d** is correct. This means there is a one-quarter cycle of the function between 0 and 2π (360°).

50. b. (Lesson 20) The common ratio of the sequence is 2. The first element is 3. Use the formula $a_n = ar^{(n-1)}$. Evaluate $a_n = 3(2^{11}) = 6,144$. If your answer is **a**, you gave the partial sum of the five elements shown in the sequence. Choice **c** is incorrect; this is the element that has the value of 12. Answer choice **d** is the partial sum of the first 12 elements of the sequence shown, not the 12th element.

ADDITIONAL ONLINE PRACTICE

Whether you need help building basic skills or preparing for an exam, visit LearningExpress Practice Center! Using the code below, you'll be able to access additional online practice. This online practice will also provide you with:

Immediate scoring
Detailed answer explanations
A customized diagnostic report that will assess your skills and focus your study

Log into the LearningExpress Practice Center by using the URL: **www.learnatest.com/practice**

This is your Access Code: **9117**

Follow the steps online to redeem your access code. After you've used your access code to register with the site, you will be prompted to create a username and password. For easy reference, record them here:

Username: _____   **Password:** _____

If you have any questions or problems, please contact LearningExpress customer service at 1-800-295-9556 ext. 2, or e-mail us at **customerservice@learningexpressllc.com**